Geometric and Discrete Path Planning for Interactive Virtual Worlds

Synthesis Lectures on Visual Computing
Computer Graphics, Animation, Computational
Photography, and Imaging

Editor

Brian A. Barsky, *University of California, Berkeley*

This series presents lectures on research and development in visual computing for an audience of professional developers, researchers and advanced students. Topics of interest include computational photography, animation, visualization, special effects, game design, image techniques, computational geometry, modeling, rendering, and others of interest to the visual computing system developer or researcher.

Geometric and Discrete Path Planning for Interactive Virtual Worlds
Marcelo Kallmann and Mubbasir Kapadia
2016

An Introduction to Verification of Visualization Techniques
Tiago Etiene, Robert M. Kirby, and Cláudio T. Silva
2015

Virtual Crowds: Steps Toward Behavioral Realism
Mubbasir Kapadia, Nuria Pelechano, Jan Allbeck, and Norm Badler
2015

Finite Element Method Simulation of 3D Deformable Solids
Eftychios Sifakis and Jernej Barbič
2015

Efficient Quadrature Rules for Illumination Integrals: From Quasi Monte Carlo to Bayesian Monte Carlo
Ricardo Marques, Christian Bouville, Luís Paulo Santos, and Kadi Bouatouch
2015

Real-Time Massive Model Rendering
Sung-eui Yoon, Enrico Gobbetti, David Kasik, and Dinesh Manocha
2008

High Dynamic Range Video
Karol Myszkowski, Rafal Mantiuk, and Grzegorz Krawczyk
2008

GPU-Based Techniques for Global Illumination Effects
László Szirmay-Kalos, László Szécsi, and Mateu Sbert
2008

High Dynamic Range Image Reconstruction
Asla M. Sá, Paulo Cezar Carvalho, and Luiz Velho
2008

High Fidelity Haptic Rendering
Miguel A. Otaduy and Ming C. Lin
2006

A Blossoming Development of Splines
Stephen Mann
2006

Geometric and Discrete Path Planning for Interactive Virtual Worlds

Marcelo Kallmann and Mubbasir Kapadia

ISBN: 978-3-031-01460-4   paperback
ISBN: 978-3-031-02588-4   ebook

DOI 10.1007/978-3-031-02588-4

A Publication in the Springer series

SYNTHESIS LECTURES ON VISUAL COMPUTING: COMPUTER GRAPHICS, ANIMATION, COMPUTATIONAL PHOTOGRAPHY, AND IMAGING

Lecture #23

Series Editor: Brian A. Barsky, *University of California, Berkeley*

Series ISSN

Print 2469-4215   Electronic 2469-4223

Geometric and Discrete Path Planning for Interactive Virtual Worlds

Marcelo Kallmann
University of California, Merced

Mubbasir Kapadia
Rutgers University

SYNTHESIS LECTURES ON VISUAL COMPUTING: COMPUTER GRAPHICS, ANIMATION, COMPUTATIONAL PHOTOGRAPHY, AND IMAGING #23

ABSTRACT

Path planning and navigation are indispensable components for controlling autonomous agents in interactive virtual worlds. Given the growing demands on the size and complexity of modern virtual worlds, a number of new techniques have been developed for achieving intelligent navigation for the next generation of interactive multi-agent simulations. This book reviews the evolution of several related techniques, starting from classical planning and computational geometry techniques and then gradually moving toward more advanced topics with focus on recent developments from the work of the authors. The covered topics range from discrete search and geometric representations to planning under different types of constraints and harnessing the power of graphics hardware in order to address Euclidean shortest paths and discrete search for multiple agents under limited time budgets. The use of planning algorithms beyond path planning is also discussed in the areas of crowd animation and whole-body motion planning for virtual characters.

KEYWORDS

path planning, Euclidean shortest paths, shortest path maps, navigation meshes, discrete search, anytime and dynamic search, multi-agent planning, whole-body motion planning, GPU search techniques

Contents

PART IV   Planning Techniques for Character Animation 127

8   Dynamic Planning of Footstep Trajectories for Crowd Simulation 129

9   Planning using Multiple Domains of Control 139

Preface

Recent research in discrete search and applied computational geometry has substantially pushed the frontiers of path planning and navigation in real-time virtual worlds. As simulated virtual worlds become more and more complex, it is expected that new techniques will continue to be developed in a number of different directions. Improved representations for navigation meshes, parallelization of search algorithms, techniques exploiting the use of graphics hardware, integration with higher-level planning layers, and integration of a variety of constraints and semantic information are example topics covered in this book which have received significant attention from research communities in artificial intelligence, robotics, computer animation, and computer games.

The text in this book originated from the SIGGRAPH 2014 course the authors offered in Vancouver, Canada [Kallmann and Kapadia, 2014]. The original material has been expanded in several ways. Discrete search techniques are explained in greater detail in order to offer the reader a concrete understanding of the basic search methods, including practical information useful for efficient implementations. Most of the other covered topics have also been significantly extended with additional techniques and examples. Overall, the chapters review relevant techniques in planning and geometric representations, with a focus on the recent work of the authors in the several related areas.

Most of the topics addressed in this book are not presented exhaustively. While some of the basic topics are covered in enough detail in order to provide a clear understanding of the basic underlying methods, several of the more advanced topics are presented as an overview of selected topics. This book provides a summarized view of recent methods in the literature, and more in particular a sample of research topics being addressed in the area. Enough information is provided for beginners to understand the basic methods well, while also presenting a good sample of challenging advanced topics to motivate readers to go beyond the basic methods. Several of the covered topics are not available in textbooks and citations to the relevant research publications are given. Motivated readers will therefore know where to look when going deeper into a preferred topic.

This book provides a presentation of selected planning techniques important for achieving autonomous navigation and intelligent motions; however, it does not cover local collision avoidance techniques which are important for agents to be able to continuously adapt their plans in dynamic environments. For comprehensive reviews on collision avoidance methods we refer the reader to other texts, such as [Kapadia et al., 2015b, Pelechano et al., 2008, Thalmann and Musse, 2007].

The target readers of this book are students, researchers, and developers with some background in computer graphics or interactive virtual worlds. While most of the covered topics find direct use in applications, some of the topics, such as using sampling-based planning for character animation, are not yet popular in practice. These topics were, however, selected in order to illustrate and stimulate new directions of research in computer animation. We expect this book to contribute to consolidating the multidisciplinary area of planning methods for interactive virtual worlds, and to provide a resource useful to practitioners and researchers in this area.

STRUCTURE OF THE BOOK

The chapters of this book start by covering classical discrete search and computational geometry techniques and then gradually cover more advanced planning topics with focus on recent developments from the work of the authors. The book is divided into four parts.

Part I (Marcelo Kallmann) describes, in the first two chapters, the foundational approaches for environment representation and discrete search methods. In Chapter 1, the main approaches for representation and discretization of planar environments are discussed with respect to their use by path search algorithms. Chapter 2 covers foundational discrete search methods that can operate in these representations. Several examples, algorithms, implementation details, and extensions are discussed.

Part II (Marcelo Kallmann) reviews geometric algorithms and representations in two chapters. Chapter 3 presents the basic methods and representations for computing Euclidean shortest paths in the plane. Chapter 4 reviews geometric representations encoding clearance information and discusses properties and approaches for navigation meshes, which are becoming the main industry approach for supporting efficient path planning in virtual worlds. The focus in Chapter 4 is on reviewing the main computational approaches and their properties, and not on reviewing the numerous implementations developed in the industry.

Part III (Mubbasir Kapadia) is composed of three chapters which extend classical search methods to meet real-time constraints, handle dynamic environments, and harness the benefits of massive parallelization in order to scale to large, complex, and dynamic environments. Chapter 5 provides an overview of extended approaches for real-time planning such as parallelization and hierarchical abstractions, and then extends the classical A* algorithm to meet real-time constraints and efficiently repair solutions in dynamic environments while preserving completeness and optimality guarantees. Chapter 6 integrates the modeling of spatial constraints that can influence path calculations. Chapter 7 presents techniques for using graphics hardware to further scale these approaches to handle large and complex environments as well as multiple planning agents.

Part IV presents, in three chapters, an overview of planning techniques extended to additional topics in character animation. Chapters 8 and 9 (Mubbasir Kapadia) discuss multi-agent navigation and crowd dynamics using different domain representations including triangulations, grids, and bio-mechanically based locomotion models for footstep selection. Chapter 10 (Marcelo

Kallmann) presents an overview of planning techniques for whole-body character motion synthesis. It discusses planning with motion capture data and sampling-based algorithms for high-dimensional planning of manipulation motions and locomotion integrated with upper-body actions.

While most of the topics are not covered in a detail level equivalent to a textbook, this book can be used as an introduction to path planning methods for computer graphics applications, for example as a complementary text in classes related to computer graphics, computer animation, and computer games. Specifically, the basic discrete search methods covered in Part I and some of the topics covered in Part II are presented in enough detail to allow students to implement them in projects. In classes related to artificial intelligence and motion planning, this book can be used to provide an overview of the representations and types of problems that appear in computer games and animated virtual environments. This can lead to ideas for project topics and to compelling motivation for learning the underlying algorithms.

Marcelo Kallmann and Mubbasir Kapadia
December 2015

Acknowledgments

This volume was made possible through the contributions of several collaborators and funding sources. The authors wish to gratefully acknowledge their role in furthering the emerging area of planning algorithms for virtual worlds.

Contributors. This book includes important contributions from the following researchers (in alphabetical order): Alejandro Beacco, Alexander Shoulson, Carlo Camporesi, Cory D. Boatright, David Huang, Francisco Garcia, Glenn Reinman, Kai Ninomiya, Mentar Mahmudi, Petros Faloutsos, Shawn Singh, and Vivek Reddy. We would also like to thank Nathan Sturtevant for the valuable suggestions during the early phases of this text.

Funding Support. A portion of the presented results was sponsored by the US National Science Foundation (NSF award IIS-0915665) and by the Army Research Laboratory (Cooperative Agreement #W911NF-10-2-0016.) The views and conclusions contained in this document are those of the authors and should not be interpreted as representing the official policies, either expressed or implied, of the Army Research Laboratory or the U.S. Government. The U.S. Government is authorized to reproduce and distribute reprints for government purposes notwithstanding any copyright notation herein. We also gratefully acknowledge Intel Corp. for their generous support through equipment and grants, and as well Autodesk, Smith Micro, and Unity for their software support.

Marcelo Kallmann and Mubbasir Kapadia
December 2015

PART I

Introduction to Discrete Search

CHAPTER 1

Basic Approaches for World Representation

Before addressing any path planning problem a suitable world representation has to be chosen. Representing a virtual environment for path planning involves designing a decomposition of the navigable space into a discrete set of connected elements such as points, lines, cells, and/or polygons. These connected elements are then represented as a graph on which discrete graph search algorithms can be applied for solving path planning queries. While there are methods that do not rely on pre-computed representations, often called *single-query* methods, the vast majority of path planning algorithms for navigation applications first build a graph representation of the environment to then rely on a discrete graph search algorithm to answer path queries.

The different ways in which a virtual world can be represented will directly influence the results obtained, the resources needed, and the path search performance. The needs of the application at hand should therefore be taken into account, such as: storage and memory space requirements, acceptable planning times, complexity of implementation, desired path properties, etc. It is common for a virtual world to rely on a representation that is specific to path planning and independent from all other modules. For example, rendering the virtual world will typically rely on a scene representation that is independent from the representation supporting path planning.

The basic approaches for representing worlds for path planning can be classified in three main classes: (a) regular grids, (b) polygonal decompositions or navigation meshes, and (c) roadmaps or waypoint graphs. This chapter describes and compares these approaches. For simplicity, the exposition in this chapter considers the representation of planar environments. Extensions for addressing 3D environments are discussed in Section 1.5.

1.1 GRIDS

Grids constitute the most straightforward way to represent an environment for path planning. The basic idea is to overlay a regular grid structure on top of the environment, and then label each cell of the grid according to the portion of the environment that lies inside the cell. If a cell contains portions of different types, the type with larger area inside the cell is typically chosen to label that cell. For example, the grid representation shown in Figure 1.1 has three types of cells which are marked with different colors: free cells are marked as gray, water cells are marked as blue, and obstacle cells are marked as brown. While the presented example is based on rectangular cells, other types of regular decompositions can also be employed. For example, hexagonal cells are also popular for representing game environments.

Because a grid is a regular decomposition, the boundaries of the cells will most often not conform to the original boundaries of the environment. The grid resolution dictates how well boundaries are represented and the resolution is controlled by the size chosen for each cell. The smaller the cells are the more precise the environment is represented. However, if the cells are too small the total number of cells will quickly become too large. In contrast, if cells are too large the regions of the environment may not be appropriately represented.

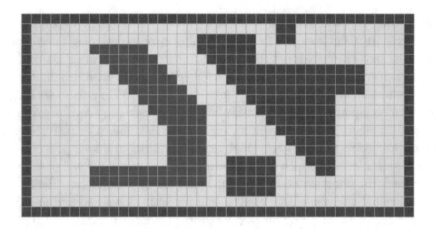

Figure 1.1:   Example of a grid representation with 557 traversable cells.

The most important parameter in a grid representation is therefore the cell size. The grid shown in Figure 1.1 has a total of 470 free cells, 87 water cells, and 243 obstacle cells. An important number to consider is the total number of cells that are allowed to be traversed by a path. In our example we allow a path to traverse free regions and water regions, what gives us a total of 557 cells that might be explored by a graph search algorithm.

A grid can be converted to a graph by connecting each traversable cell to its four neighbors on the horizontal and vertical directions. Figure 1.2 illustrates the obtained graph in our example. Another option is to consider that the grid is 8-connected, meaning that each cell is connected to all its 8 neighbors with graph edges, enabling the path search to consider diagonal moves in addition to cardinal moves and resulting in better path solutions. The tradeoff is that the number of edges per node increases, slowing down the overall path search. A comparison example of both approaches is later discussed in the next chapter.

One advantage of grid-based representations is that the environment becomes covered by a regular distribution of cells and thus it becomes straightforward to represent different traversal costs across the traversable areas. This is achieved by associating each edge in the graph with a traversal cost representing the difficulty of traversing the terrain between the two cells connected by the edge. In our example, edges that traverse water regions could be associated, for example, with a cost two times more expensive than the cost associated with an edge traversing a regular free region. Edges that cross two different types of terrain may have an in-between cost. The graph search algorithms that we will later use will take into account the different edge costs in order to compute a path of minimal overall cost in the graph.

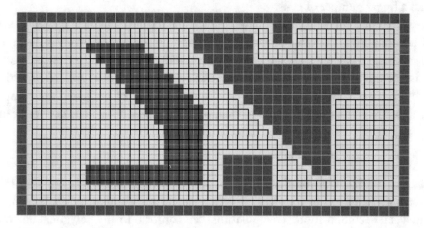

Figure 1.2:   Graph search algorithms will compute paths in a grid by searching a connectivity graph such as the one shown in black. The graph connects with edges all pairs of traversable cells which are neighbors in the vertical and horizontal directions.

Although graph search algorithms will compute paths of minimum cost, it is important to observe that a minimum cost path in such a graph will most often not correspond to the minimal cost path in the continuous environment the graph represents. This is also the case if all edge costs are the same, and also holds for all representations discussed in this chapter. This point will become clearer in the second part of this book when the Euclidean Shortest Path problem will be discussed.

1.2   POLYGONAL DECOMPOSITIONS AND NAVIGATION MESHES

Polygonal decompositions can exactly conform to the boundaries of a given polygonal environment and will usually do so with much less cells than grid representations. In order to achieve such flexibility, the cells of a polygonal decomposition will be simple polygons of varied size and type.

Among the several approaches for generating polygonal decompositions of a given environment, trapezoidal decompositions [Chazelle, 1987, de Berg et al., 2008] are among the simplest ones. These decompositions can be built by placing vertical (or horizontal) lines starting at each vertex of the environment and ending at the first boundary intersection. The resulting segments will form a trapezoidal decomposition of the environment. The number of polygonal cells in the obtained decomposition will be linear with respect to the number of vertices n in the original polygonal boundaries. In other words, the number of cells is $\mathcal{O}(n)$, an important property that several polygonal decompositions can ensure. Trapezoidal decompositions are, however, not popular in practice because they generate long and skinny polygonal cells without useful properties other than a linear number of cells.

Polygonal decompositions can have different forms and a natural goal is to seek for a decomposition that offers desirable properties. For example, an optimal convex decomposition will have the smallest possible number of convex polygonal cells [Chazelle, 1987]. Another criterion is to minimize the total length of all edges in the subdivision. These and other optimality criteria are the subject of active research in computational geometry, but given the complexity of the involved algorithms their practical use in path planning applications has not yet been explored.

1.2.1   TRIANGULATIONS

Triangulations represent one particular type of decomposition that is very popular. Triangulations decompose a given region in triangles, which represent the simplest possible planar element to be used in a cell decomposition. Triangulations can conform to any polygonal environment with $\mathcal{O}(n)$ triangular cells, and they can be built in different ways according to desired properties.

One particular type of triangulation that is very popular is the Delaunay Triangulation (DT), which maximizes the minimum internal angle of its triangles. In our case, since we are interested in subdivisions that conform to given boundaries, our interest is on the Constrained Delaunay Triangulation (CDT). A CDT ensures that given boundary edges of an environment will also be edges of the triangulation, while maintaining the DT property as much as possible. Several efficient algorithms are available for building CDTs, and they also produce a cell decomposition of the navigable space in $\mathcal{O}(n)$ cells. Figure 1.3 illustrates the CDT of our test environment. In this chapter we present CDTs as an example of a polygonal cell decomposition representation. A more formal definition and detailed discussion will be presented in Chapter 4.

The triangulation shown in Figure 1.3 produces 34 free cells, 6 water cells, and 12 obstacle cells. These numbers are clearly much lower than the ones obtained with the corresponding grid decomposition (Figure 1.1). This is as expected since the number of cells in a CDT is $\mathcal{O}(n)$. Linear size decompositions are important because they guarantee a low number of cells, consequently leading to fast path searches.

Note that the user has some flexibility in deciding how many vertices and edges to use when describing the polygonal boundaries of the environment. Polygonal boundaries may be selected to simplify or further refine the original specifications. For example, in Figure 1.3 there is one redundant vertex subdividing a horizontal boundary (in the top-right horizontal edge of the right-most obstacle). If that vertex is removed, the original triangles connected to the removed vertex would be replaced by fewer triangles of larger size, but the boundary would continue to be exactly represented. Such redundant vertices, often called *Steiner vertices*, are sometimes added to

Figure 1.3: The shown CDT is one example of a polygonal decomposition that exactly conforms to the boundaries of the given environment. The resulting decomposition has $\mathcal{O}(n)$ triangular cells. In this example $n = 30$ and there are a total of 40 traversable triangular cells.

a triangulation in order to refine the decomposition for a given purpose, such as for ensuring a desired property. We will see in Chapter 4 how this technique can be used in a CDT in order to efficiently encode clearance information in a *Local Clearance Triangulation*.

1.2.2 EULER'S POLYHEDRAL FORMULA

Another useful property of planar polygonal decompositions is that they have the same combinatorial properties as 3D polyhedra. This happens because they are basically the same structures. Consider a mapping to the 3D space where the vertices on the border of the polygonal decomposition are lowered along the Z axis, while the other vertices remain on their original XY plane, as shown in Figure 1.4. Note that the original decomposition is considered to have a *back face* that is the one that gets mapped to the bottom face of the obtained 3D polyhedron. Another possible mapping is to project the 2D points to the surface of a spherical cap placed over the 2D decomposition, in this case producing a polyhedron without co-planar 3D faces. In both cases, after the mapping, polyhedron combinatorics such as the Euler's characteristic formula can be applied.

The classical Euler characteristic formula for polyhedra is defined as: $V - E + F = \chi$, where V is the number of vertices, E is the number of edges, F is the number of faces, and χ is the Euler characteristic. In our case, because our 2D environment is mapped to a 3D polyhedron that is always topologically equivalent to a sphere, the Euler characteristic will always be 2 and the formula becomes $V - E + F = 2$, which is also known as the polyhedral formula.

In a triangulation all *faces* are triangles and the polyhedral formula can be written independently of the number of edges. We start by counting the number of edges as a function of the number of triangles t and the number of edges b in the border of the triangulation. The number

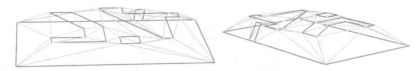

Figure 1.4: Two different views of the resulting mapping from the planar decomposition shown in the previous figure to a 3D polyhedron. The mapping is obtained by lowering the border vertices along the orthogonal axis of the subdivision plane.

of edges will be $(3t/2) + (b/2)$. The first term $3t/2$ accounts for 3 edges per triangle, divided by 2 because each edge is shared by 2 triangles. However, in doing so, we have also accounted for half of the border edges, and the second term $b/2$ adds the remaining amount to correctly account for the boundary edges. We can then apply the polyhedron formula, where $V = n$, $E = (3t/2) + (b/2)$, and $F = t + 1$, where the number of faces also accounts for the *back face* in the 3D polyhedral mapping (see Figure 1.4). The polyhedral formula becomes $n - [(3t/2) + (b/2)] + (t + 1) = 2$, which can be simplified to $t = 2n - 2 - b$. This formula proves that the number of triangular cells in the decomposition is $\mathcal{O}(n)$, and it shows that in most cases the number of triangles will be approximately 2 times the number of vertices. It is straightforward to verify the formula in our environment of Figure 1.3, where $t = 52$, $n = 30$, and $b = 6$.

Once the polygonal decomposition is computed, the adjacency of the polygonal cells can be used to build a graph on which graph search algorithms can search for paths. The most straightforward way to build such a graph is to define the nodes of the graph to be the centers of the navigable triangles, and to define the edges to connect pairs of nodes that are on adjacent triangles. Figure 1.5 illustrates the graph obtained.

Figure 1.5: Adjacency graph connecting the centers of adjacent triangles that are navigable.

Another possible way to create such an adjacency graph is to use as connection points the midpoints of the traversable edges. The use of edge midpoints have shown to generate better graphs in generic environments and additional options are possible for deriving a triangulation adjacency graph suitable for path search [Kallmann, 2010]. Note that the adjacency graph usually does not need to be computed explicitly. Instead, during the execution of the graph search algorithm, the adjacency information can be directly extracted from the data structure representing the cell decomposition.

Finally, the edges of the graph are also associated with traversal costs. For an edge that traverses two triangles of the same type the cost will typically be the length of the edge multiplied by the traversal cost of the respective terrain type. For edges that cross two different types of terrain, the cost should be the sum of the costs independently computed for each sub-edge lying on a different terrain type.

1.2.3 NAVIGATION MESHES

The term *navigation mesh* was coined by the computer game community and is used to refer to any mesh that describes navigable surfaces for path planning and other navigation queries. A mesh here refers to a collection of polygons that together describe a navigable polygonal surface. A formal polygonal decomposition is expected to have the property that its cells do not overlap and that the union of all cells completely cover the area being represented. While a navigation mesh is considered here to be equivalent to a polygonal decomposition, the term is commonly employed without any guarantees on properties. Chapter 4 includes an overview of the main topics related to designing navigation meshes, from extracting the navigable surfaces in 3D environments to the underlying geometric representations and properties.

1.3   ROADMAPS AND WAYPOINT GRAPHS

A common point in the approaches discussed so far is that they first represent the navigable space with some type of cell decomposition to then derive an adjacency graph that can be used for path search. While the cell decomposition is important in order to define a graph that captures the navigable space of the environment well, graphs can also be built independently of a cell decomposition.

1.3.1 SAMPLING-BASED ROADMAPS

One popular method in robotics is to randomly sample points in the environment in order to construct a graph for path planning. For example, the *Probablistic Roadmap* (PRM) [Kavraki et al., 1996] approach starts by computing a set of random points determined to be in the navigable portion of the environment. These points become the nodes of the graph and the edges are determined by connecting pairs of nodes that are close to each other, if the connection represents a collision-free segment. The result is a PRM graph.

One advantage of PRMs and other sampling-based methods is that they are simple enough to be applied for path planning in high dimensions. Several other sampling-based methods have been developed for high-dimensional planning, including single-query methods that grow trees (instead of graphs) rooted at the query points [Kuffner and LaValle, 2000]. These methods will be reviewed in Chapter 10.

One drawback of sampling-based methods is that it is difficult to know when the generated graph has captured all the *corridors* of the given environment, and therefore it becomes difficult to determine how many samples are enough to correctly answer a planning query. Besides simplicity of implementation, roadmaps based on random sampling do not offer a direct advantage for 2D path planning in planar surfaces. The usually slower construction time, the potentially high number of nodes, and the difficulty to determine when to stop the random sampling are limitations that can be avoided in 2D by relying on cell decompositions.

1.3.2   WAYPOINT GRAPHS

Another approach to generate graphs, or roadmaps, is to design them by hand. This may sound impractical, however, it is a common practice in the development of certain types of videogames. Environments in games are often designed from scratch by artists, who spend significant time modeling the environment and annotating a variety of information. Several editing tools are developed to assist designers in developing the environments, which are often organized by levels. Editing tools are commonly called as level editors. Annotating a roadmap graph for path planning can be easily accomplished by clicking in the environment where the nodes of the roadmap should be placed. The obtained graph is usually called a *waypoint graph* by the computer game community. A roadmap and a waypoint graph are essentially different names for the same structure: a graph representing the navigable areas of the environment. Figure 1.6 illustrates a possible roadmap designed by hand for our test environment.

The hand-designed graph in Figure 1.6 is quite similar to the graph in Figure 1.5; however, since it was designed by hand, the graph in Figure 1.6 was intuitively built to minimize unnecessary sharp angles or dangling edges.

1.4   COMPARISON

From the examples given in this chapter it is possible to observe that the chosen representation will have a direct impact on the graphs that are generated for path search. Figure 1.7 provides a side-by-side comparison.

In general, grids provide a dense representation of the navigable spaces and therefore are suitable to represent information such as terrain costs across the environment. Polygonal cell decompositions will generate compact graphs suitable for fast path search but can only represent information per-polygon and a same polygon may cover a large portion of the environment. Roadmaps (or waypoint graphs), when not associated with a cell decomposition, require a method

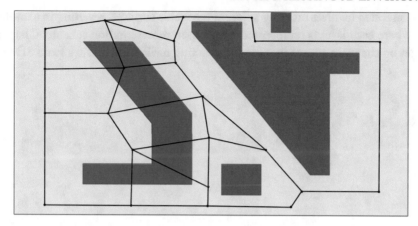

Figure 1.6: Example of a hand-designed roadmap (or waypoint) graph.

Figure 1.7: Graphs obtained from the different examples discussed in this chapter. In left to right order: grid adjacency graph, CDT adjacency graph, and hand-designed roadmap.

to determine the location of the nodes of the graph. Such graphs can alternatively be designed by hand and fine-tuned as needed.

In many applications the ability to represent spatial information at fine resolutions is not important and in these cases the focus becomes on computing paths as fast as possible. In these cases polygonal cell decompositions with guaranteed $\mathcal{O}(n)$ cells become the best option. The popularity of navigation meshes in recent years reflects the fact that polygonal decompositions can provide efficient representations for path planning.

1.5 ADDRESSING 3D ENVIRONMENTS

A number of additional steps have to be taken into account in order to build navigation structures for 3D environments. First of all, the navigation capabilities of the agents to be simulated have to be clearly specified. Based on these capabilities, surfaces that are acceptable for navigation can be extracted from an input 3D world and finally connected to each other according to the chosen planar representations. The typical approach is to analyze the virtual world globally, with the use of a volumetric decomposition of the whole space occupied by the scene.

The final step involves merging adjacent navigable surfaces, resulting in a multi-layer representation where each layer is represented in a chosen representation scheme. Chapter 4 includes a discussion on the main approaches for constructing navigation meshes from 3D environments.

CHAPTER 2

Discrete Search Algorithms

As we have seen in the previous chapter, given a planar environment, different methods can be used to represent the navigable space of the environment with a graph. The process transforms the continuous path planning problem into a discrete planning problem. For each node of the graph, there are only a discrete number of choices to decide in which direction to move next.

Discrete search algorithms do not actually require an explicit graph to exist. It is enough to have a state transition function f that returns the next state for every possible action to be taken from the current state. Let u denote an action and x the current state. The next state obtained after applying action u will be $x' = f(x, u)$. This notation is more generic since it does not require the entire state graph of the problem to be explicitly available, it only requires the moves between states to be known. Function f can represent the possible moves of an agent and as well of any system that is described by states. For example, when solving a Rubik's cube puzzle, the transition function f is well defined because all the discrete number of possible rotation moves to change the cube from the current state to any adjacent state are known. However, even though a solution can be always found with at most 20 moves, it would be practically impossible to represent all the states of the cube explicitly in a graph because the number of states is just too large (43 quintillion! [Rokicki et al., 2013])

In the case of navigation in virtual worlds, the navigable space is often entirely known in advance and usually can be fully represented by a graph. If the graph cannot be fully represented, partial graphs can be used to represent the current portion of the navigable space being processed. This happens, for example, in environments that are too large to be loaded entirely in memory at once. We consider in this chapter that the navigable space of interest (partial or full) is completely available and its connectivity graph is readily available for graph search.

While there are a number of discrete search algorithms available [Cormen et al., 2009, LaValle, 2006], this chapter focuses on the two most notable ones in the scope of this book: Dijkstra and A* algorithms.

2.1 DIJKSTRA ALGORITHM

Edsger Wybe Dijkstra introduced his algorithm in a 3-page 1959 paper addressing two problems: to compute the minimum length tree passing by all nodes, and to compute the shortest path between two nodes [Dijkstra, 1959]. The concept behind both solutions is the same, it is based on expanding a frontier of equal distance from a source node, expanding outwards from it, step by step, until all nodes are processed or until a goal node is reached. The algorithm is elegant and effective, and when heuristics are not known or not possible, it remains the algorithm of choice for many planning problems.

2.1.1   ALGORITHM

An efficient implementation of Dijkstra's shortest path algorithm will rely on a priority queue able to efficiently maintain the nodes in the expansion front sorted according to their *cost-to-come* values computed so far. The cost-to-come value of a node n is the cost of the lowest-cost path from the source node to n. The sorting ensures that nodes are expanded in order, taking into account edge traversal costs, until the goal node is reached and the shortest path to it can be determined.

If the navigation graph is from a uniform terrain with constant traversal costs, edge costs will be the distance between the end points of the edges and a lowest-cost solution will be a shortest path. In non-uniform terrains the cost of traversing one edge can be estimated by multiplying the edge length by the terrain cost associated with the edge. The final edge costs considered by the algorithm are restricted to be positive, therefore restricting terrain costs to be always positive. If a terrain is modeled with a negative cost it is also possible to elevate the costs of all the edges in the terrain such that the smallest cost becomes positive and the relative costs between the different areas in the terrain are preserved. Negative costs can lead to negative-weight cycles and specialized algorithms would have to be employed, such as the Bellman-Ford algorithm [Cormen et al., 2009]. Usual navigation problems do not have a need to address negative costs and so the fact that the algorithms discussed in this chapter require positive costs does not a represent a real restriction.

One important factor in the overall performance of Dijkstra's algorithm is an efficient priority queue implementation. Section 2.1.4 analyzes worst-case running times for typical implementations of priority queues, and will describe simple implementation approaches that are also fast for the typical cases encountered in navigation graphs.

In the general case, a priority queue Q has to process three key operations: insertion, removal, and priority decrease of nodes. The priority value of a node is its cost-to-come value. Let n be a node being inserted in Q with associated priority cost c. The insertion function $Q.insert(n, c)$ will store node n with priority c in Q. Inserted nodes should be kept well organized in Q such that, at any time, a removal call $Q.remove()$ will efficiently remove and return the node of minimal cost among all nodes in Q. Function $Q.decrease(n, c)$ will decrease the priority of n, a node already in Q, to the new priority c.

Algorithm 1 provides a complete implementation of Dijkstra's algorithm. It relies on priority queue Q to sort the nodes in the expansion frontier according to their current cost-to-come costs, which are stored at the nodes and retrieved with function $g(n)$. The algorithm receives as input the start node s and the goal target node t. In practice, path queries are in the Euclidean plane and will be specified by points and not graph nodes. In this case the input points are converted to their respective nearest graph nodes s and t in the graph, such that **Dijkstra**(s, t) can be then called.

Algorithm 1 - Dijkstra Algorithm for Shortest Paths

Input: source node s and goal node t.

Output: shortest path from s to t, or null path if it does not exist.

  1: **Dijkstra**(s, t)
  2: Initialize Q with $(s, 0)$, set $g(s)$ to be 0, and mark s as visited;
  3: **while** (Q not empty) **do**
  4:    $v \leftarrow Q.remove()$;
  5:    **if** ($v = t$) **return** reconstructed branch from v to s;
  6:    **for each** (neighbors n of v) **do**
  7:      **if** (n not visited **or** $g(n) > g(v) + c(v, n)$) **then**
  8:        Set the parent of n to be v;
  9:        Set $g(n)$ to be $g(v) + c(v, n)$;
10:        **if** (n visited) $Q.decrease(n, g(n))$; **else** $Q.insert(n, g(n))$;
11:        Mark n as visited, if not already visited;
12:      **end if**
13:    **end for**
14: **end while**
15: **return** null path;

Algorithm 1 starts by initializing Q as empty and then inserting s in Q with associated cost 0. The cost-to-come $g(s)$ of node s is set to 0 and s is also marked as visited. Nodes are marked in order to distinguish nodes that are being processed for the first time (non-marked nodes) from nodes that have been already processed and associated with a cost.

In line 3 the algorithm enters a loop that continues while there are nodes in Q to be expanded. Minimum cost node v is then removed from Q and processed. If v is the goal node, the algorithm has reached the goal and can return the branch of the search from v to s. If the goal is not reached, then each neighbor n of v is processed. Node n will be inserted in Q if it is being visited for the first time, or if it is being visited by a shorter path than before (line 7 of the algorithm). This latter test is verified with $g(n) > g(v) + c(v, n)$, where $g()$ retrieves the current cost-to-come associated with already-visited nodes, and $c(v, n)$ retrieves the cost to move from v to n, which combines the distance from v to n and any other traversal property stored in the edge connecting v to n.

As the search progresses, each node being inserted in Q keeps a pointer to the node found to be its parent along the lowest-cost path to s. Every time n is inserted, its parent node is set to be v (line 8). Setting parent nodes is needed for allowing the retrieval of the solution branch as required in line 5, which is accomplished by tracing back all parent nodes, from the goal node back to the source node. Line 9 computes the cost-to-come to be associated with n, n is then inserted or updated in Q with the new cost, and finally marked as visited if needed.

The process continues by removing the next minimum cost node v from Q. If Q becomes empty, it means that all nodes connected by a path to s have been visited and t has not been reached. This means that s is in a disconnected component of the graph and no path from s to t exists. In this case, a null path is returned.

Textbooks in data structures will often present the Dijkstra algorithm by first inserting in Q all nodes in the graph with an infinite cost, and then costs are gradually reduced as the algorithm progresses. Cost reduction is an operation called *relax* [Cormen et al., 2009], which is equivalent to lines 7–9 in Algorithm 1. Here we rely on marking nodes to distinguish non-processed nodes from already visited ones, eliminating the need to insert all nodes in Q, and having Q to only contain the active frontier being expanded. Once a node is removed from Q its cost will be final because all its predecessors will have been already processed.

2.1.2   STEP-BY-STEP EXAMPLE

A step-by-step example of Dijkstra's algorithm is illustrated in Figure 2.1. The top-left diagram illustrates Dijkstra's algorithm right after initialization, where the source node s is marked as visited (node s appears with a darker shading), $g(s)$ is set to 0, and Q is initialized with node s and its associated cost 0. The next diagrams are numbered by algorithm iteration, each one illustrating the state of the algorithm after completing one iteration, i.e., after completing the while loop of Algorithm 1 one time. In each diagram, the most recent node removed from Q and the neighbors that were altered by the end of the iteration are highlighted. Each node shows its current cost-to-come, and a dash character indicates that no value has been determined yet.

The search starts from node s. The first iteration of the algorithm will process s, reach its neighbors r, v, and u, and place them in Q sorted according to their cost-to-come values. At iteration 2, r is removed from Q and its neighbor v is reached for the second time, triggering a decrease key operation that decreases the cost of v from 4 to 3 causing v to be upgraded to first element in Q. The decrease key operation reflects the fact that, even though more edges are traversed, v is reached with lower cost when passing by r instead of coming directly from s. Iteration 3 then processes node v, when its neighbors w and t are reached and inserted in Q with their respective costs. Although t is the target node the algorithm does not end until t is removed as the lowest cost element from Q, and currently t is inserted in Q as its third element with cost 6. Iteration 4 processes node u, which is also a neighbor of t, causing t to be reached with cost-to-come of 5, triggering a decrease key operation for t and placing it as first element of Q. At iteration 5 the algorithm removes the lowest cost element t from Q, detects it to be the target node, and finally returns the traversed branch from t to s as the solution path.

2.1.3   GRID EXAMPLE

Figure 2.2 illustrates Dijkstra's algorithm running on the 4-connected grid graph of the example environment used in the previous chapter. In this example there are several regions with uniform traversal costs, and as a grid, there is a relatively high number of nodes to be expanded. The

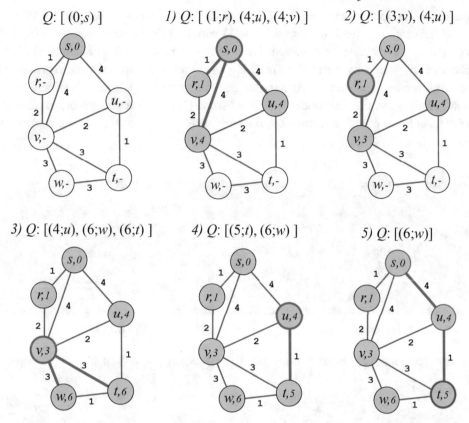

Figure 2.1: Step-by-step illustration of Dijkstra's algorithm. The top-left diagram illustrates the state of the algorithm right after initialization. The remaining diagrams are numbered by algorithm iteration and illustrate the state of the algorithm after completing the respective iteration. The solution branch is determined at iteration 5.

highlighted nodes in the figure are the nodes in Q, which are the nodes representing the current expansion frontier. The nodes in the expansion frontier have approximately equal cost to the source node. The costs are approximately equal because, since the search is discrete and performed one step at a time, the cost-to-come encoded in the nodes in Q may vary by the cost of one step. As in the previous example, the priority queue Q keeps the nodes sorted such that the lowest cost node is always expanded first, progressively expanding the frontier. For each iteration of the main loop in Algorithm 1, one node is expanded, i.e., one node is removed from Q and its neighbors are processed.

After 100 iterations it is possible to notice in Figure 2.2 that the nodes in Q maintain an exploration front that, although asymmetric, denote a frontier of approximately equal cost to the source node. The front shows more nodes expanded to the left side because that side only has low-cost traversal cells (each of traversal cost 1) while on the other directions water cells (each of traversal cost 2) have to be traversed. Later at 385 iterations it is possible to note that many branches of the search have ended by reaching the outer wall and only two corridors are being explored with 6 leaf nodes in Q. At iteration 484 the goal node is reached and added to Q, and then at iteration 486 the goal node is removed from Q, at which point it is guaranteed that no other node of lower cost exists, therefore the path leading to the goal node is an optimal solution path.

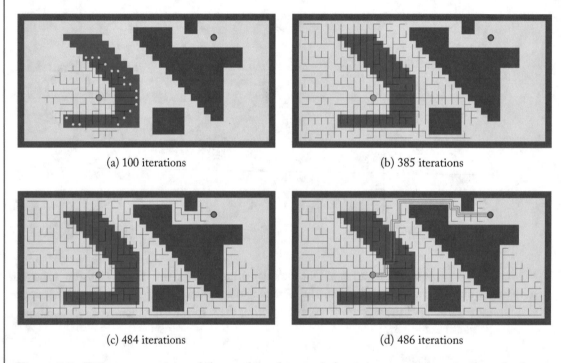

(a) 100 iterations

(b) 385 iterations

(c) 484 iterations

(d) 486 iterations

Figure 2.2: Dijkstra expansions while searching for a path from the green sphere to the red sphere. The generated search tree is shown in black and the nodes (leafs) in Q are highlighted in yellow.

As Figure 2.2 shows, graphs extracted from grid representations lead to paths with many turns and with different sets of turns producing different paths with a same overall cost. In these cases several optimal paths of same minimum cost exist, and Dijkstra's algorithm will find one of them. Path post-processing or path smoothing techniques can be applied if the quality of the obtained paths need to be improved, which is an application-dependent determination.

2.1.4   ANALYSIS

The overall running time of the Dijkstra algorithm will depend on the time taken for each operation in Q. Priority queue Q is usually implemented with a self-balancing binary search tree or with a binary min-heap [Cormen et al., 2009], which are structures that can guarantee each of the insertion, removal, and decrease operations to take $\mathcal{O}(\log n_q)$ computation time, where n_q denotes the number of elements in Q. In terms of implementation efficiency, a binary heap is often preferred because it can be implemented in an array without the need to manipulate pointers.

A binary heap however does not offer a straightforward implementation for the $decrease()$ operation. A simple implementation would need to perform a linear search of $\mathcal{O}(n_q)$ time to find the element in Q, to then perform a $\mathcal{O}(\log n_q)$ procedure to adjust its position to the new key. To prevent the linear search each node in the graph has to maintain a value indicating its current position in Q, and to continually update the value as the node position in Q changes during the progression of the algorithm. This involves significant manipulation of indices but achieves the running time of $\mathcal{O}(\log n_q)$.

Overall, the algorithm calls $insert()$ and $remove()$ once per vertex, while $decrease()$ is called up to m times, where m is the number of edges in the graph. Since $insert()$ is called once per vertex, Q will have at most n elements, where n denotes the total number of vertices in the graph. The total running time is therefore $\mathcal{O}((n + m) \log n)$, which reduces to $\mathcal{O}(m \log n)$. It is possible to achieve a running time of $\mathcal{O}(m + n \log n)$ by implementing the min-priority queue as a Fibonacci heap, which achieves an implementation of $decrease()$ in constant amortized time. This may represent a good improvement on graphs dense in edges. For example, in a graph with $m = \mathcal{O}(n^2)$ the overall worst-case Dijkstra time is reduced from $\mathcal{O}(n^2 \log n)$ to $\mathcal{O}(n^2)$. The development of Fibonacci heaps was motivated by the fact that graphs in several problems are dense and with highly irregular costs, in which case Dijkstra's algorithm will most likely perform more decrease-key operations than insertion operations. With the exception of algorithms seeking for Euclidean shortest paths using visibility graphs (Section 3.2), our navigation structures of interest will have $m = \mathcal{O}(n)$ and the need for Fibonacci heaps is not justified.

As an example, in the algorithm execution illustrated in Figure 2.2, Algorithm 1 did not need to perform any decrease-key operations. This happened because the edge density is low and the edge costs are uniform in most of the areas of the environment. There are several alternate paths that could be chosen but because they generate paths of same cost they do not trigger the test in line 8 that would lead to a decrease-key operation. As seen in the example of Figure 2.1, the decrease-key operation is however needed in order to ensure that optimal paths in the graph are found for the general case. Figure 2.3 illustrates that a decrease-key operation can happen in very simple situations with non-uniform costs, and that it is also important to test for target arrival at the right moment, when the target node is removed from Q and not when it is first reached.

A good simplification that works well for navigation graphs is to not decrease the key of already inserted nodes and instead insert a duplicate of the node in Q with the correct updated cost. This approach is shown in Algorithm 2, which only needs insertion and removal opera-

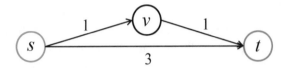

Figure 2.3: This example requires a decrease-key operation when vertex t is reached from v.

tions. The duplicate nodes in Q with non-updated values are detected at removal time (line 6) and simply not processed. During the performance of the algorithm the number of elements in Q may become larger than in the original Algorithm 1; however, the overall running time remains $\mathcal{O}(m \log n)$ and the simplification obtained by not having to maintain extra information to implement a decrease-key operation makes the queue implementation simpler and faster. Benchmark comparisons demonstrating concrete speed gains have already been reported in [Chen et al., 2007].

Algorithm 2 - Simplified Dijkstra Algorithm for Shortest Paths

Input: source node s and goal node t.
Output: shortest path from s to t, null path if it does not exist.

1: **SimplifiedDijkstra**(s, t)
2: Initialize Q with s associated with cost 0.
3: Mark s as visited;
4: **while** (Q not empty) **do**
5:     $(v, k) \leftarrow Q.remove()$;
6:     **if** ($k < g(v)$) **continue**;
7:     **if** ($v = t$) **return** reconstructed branch from v to s;
8:     **for each** (neighbors n of v) **do**
9:         **if** (n not visited **or** $g(n) > g(v) + c(v, n)$) **then**
10:            Set the parent of n to be v;
11:            Set $g(n)$ to be $g(v) + c(v, n)$;
12:            $Q.insert(n, g(n))$;
13:            Mark n as visited, if not already visited;
14:        **end if**
15:    **end for**
16: **end while**
17: **return** null path;

Another relevant consideration to make is that, in the case of the navigation graphs being considered, finding the optimal shortest path in the graph most often does not translate into finding the shortest path in the plane. Therefore, it is worth considering optimizations even if finding an optimal shortest path in the graph cannot be guaranteed. Algorithm 3 presents such an optimization. It differs from the Dijkstra algorithm by not testing for the cases that may lead to a decrease-key or duplicate node insertion operation. In this way the algorithm becomes even simpler and the number of elements in Q is reduced. The obtained paths however are not guaranteed to be optimal, and this is why Algorithm 3 is called *Forward Search*. Nevertheless, in many practical grid-based cases (as in the example of Figure 2.2) the results are not affected.

Algorithm 3 - Dijkstra-like Forward Search Algorithm

Input: source node s and goal node t.

Output: a path from s to t, null path if path does not exist.

1: **ForwardSearch**(s, t)
2: Initialize Q with s associated with cost 0.
3: Mark s as visited;
4: **while** (Q not empty) **do**
5:    $v \leftarrow Q.remove()$;
6:    **if** ($v = t$) **return** reconstructed branch from v to s;
7:    **for each** (neighbors n of v) **do**
8:      **if** (n not visited) **then**
9:        Set the parent of n to be v;
10:        Set $g(n)$ to be $g(v) + c(v, n)$;
11:        $Q.insert(n, g(n))$;
12:        Mark n as visited;
13:      **end if**
14:    **end for**
15: **end while**
16: **return** null path;

Algorithm 3 still maintains an expansion front of approximate equal cost to the source node s, but it will not treat cases where one edge move may win over past moves, as shown in Figure 2.3. Algorithm 3 will mark visited nodes that have been already processed, and all marked nodes will be inserted in Q only one time.

An efficient way to implement marking without requiring to visit all nodes for resetting a flag every time the algorithm is called is to use an integer instead of a Boolean flag. At first all integers start initialized as zero and a marking value k is initialized as 1. A node is considered marked if its integer value is equal to k. Every time marking is needed, k is incremented by one and a node is marked as needed by assigning to its integer the value of k. In this way resetting markers is reduced to incrementing k, and only when k reaches its maximum possible value that all integers in the nodes have to be re-initialized to zero and k set again to 1.

In summary, the three variations of Dijkstra's algorithm discussed in this section will run in time $\mathcal{O}(n \log n)$ for the typical navigation graph case of $m = \mathcal{O}(n)$. Algorithm 1 assumes the availability of a priority queue supporting a decrease-key operation. Algorithm 2 is of simpler implementation because it does not require decrease-key operations and it will run faster in most navigation graphs given the usually low density of edges. Algorithm 3 is even simpler and keeps less nodes in Q, but will not guarantee that optimal paths are always found.

The discussed variations cover relevant considerations to be considered when implementing priority queues for generic situations. Another approach for optimizing queue operations is to design queues with bucketed entries, an approach which has shown to provide significant speedups [Goldberg, 2001]. In particular, for the case of grid environments with uniform costs where the range of possible cost values can be relatively small, bucketed queues can be designed to practically operate in constant time.

2.2   A* ALGORITHM

The A* algorithm [Hart et al., 1968] is an extension of Dijkstra's algorithm for achieving better performance with the use of heuristics which are based on knowledge about the specific problem being solved. Because of this strategy, it belongs to a class of algorithms known as heuristic search or informed search. The general idea is to expand nodes taking into account their estimated distances to the goal. In comparison to Dijkstra's algorithm, instead of maintaining a frontier of equal cost to the source node, the frontier maintains nodes with estimated equal costs along their shortest paths to the goal, according to a heuristic function that estimates how far each node is from reaching the goal along its shortest path.

A* sorts nodes in queue Q using a cost function f composed of two parts: $f(n) = g(n) + h(n)$, where $g(n)$ is the same cost-to-come cost of Dijkstra's algorithm, and $h(n)$ is the heuristic cost that estimates the cost of the lowest-cost path from n to the goal node. While implementing an accurate estimate for function h would be difficult, the estimate just needs to be *admissible*; that is, it must not overestimate the cost to reach the goal. For navigation applications a popular simple approach is to encode in $h(n)$ the straight-line distance from n to the goal. Although this estimate is often not accurate, it is admissible, simple to implement, and already provides a useful heuristic. The heuristic cost is often referred to as the *cost-to-go* cost.

2.2.1   ALGORITHM

Algorithm 4 presents an implementation of A* based on Algorithm 1. Only one line is changed (line 10), where function $f(n) = g(n) + h(n)$ is used instead of $g(n)$ to sort the nodes in Q. This change will expand a frontier that has approximately equal lowest-possible overall cost to reach the goal. This results in a frontier where the portions nearest to the goal node t will progress faster toward t, without loosing the optimality of the solution. When the heuristic function is not used ($h(n) = 0$), the algorithm becomes a Dijkstra expansion.

Algorithm 4 - A* Algorithm for Shortest Paths

Input: source node s and goal node t.

Output: shortest path from s to t, null path if it does not exist.

1: **AStar**(s, t)
2: Initialize Q with $(s, 0)$, set $g(s)$ to be 0, and mark s as visited;
3: **while** (Q not empty) **do**
4:     $v \leftarrow Q.remove()$;
5:     **if** ($v = t$) **return** reconstructed branch from v to s;
6:     **for each** (neighbors n of v) **do**
7:         **if** (n not visited **or** $g(n) > g(v) + c(v, n)$) **then**
8:             Set the parent of n to be v;
9:             Set $g(n)$ to be $g(v) + c(v, n)$;
10:            **if** (n visited) $Q.decrease(n, g(n) + h(n))$; **else** $Q.insert(n, g(n) + h(n))$;
11:            Mark n as visited, if not already visited;
12:        **end if**
13:    **end for**
14: **end while**
15: **return** null path;

2.2.2   GRID EXAMPLE

Figure 2.4 illustrates A* running on the same grid graph example used to demonstrate Dijkstra's algorithm. The highlighted nodes in the figure are the leaf nodes in Q. At 85 iterations it is possible to notice that the nodes in Q that are closer to t have expanded much more than the ones that are on the opposite side of the expansion frontier. Basically the frontier expands faster toward the goal, and this happens even with the need to traverse the high-cost water area. Later at 285 iterations the search reaches the two corridors that will compete for the lowest-cost path to the goal. The top corridor is on a straight line direction to the goal and clearly expands much faster from this point on and at 321 iterations the search first encounters the goal node t. The algorithm terminates in the next iteration, when t is removed from Q, determining that no other path of lower cost is possible. A* reached the solution in 322 iterations while Dijkstra's algorithm took 486 iterations.

2.2.3   ANALYSIS

In terms of worst-case run time complexity A* expands nodes equivalently to Dijkstra's algorithm and therefore the analysis presented in Section 2.1.4 is also valid for A*. The simplifications discussed in that section can also be applied for the A* algorithm, meaning that equivalent versions not requiring the decrease-key operation can also be implemented for A*. The only difference is that the A* versions will add $h(n)$ to the key value sorting the nodes in Q.

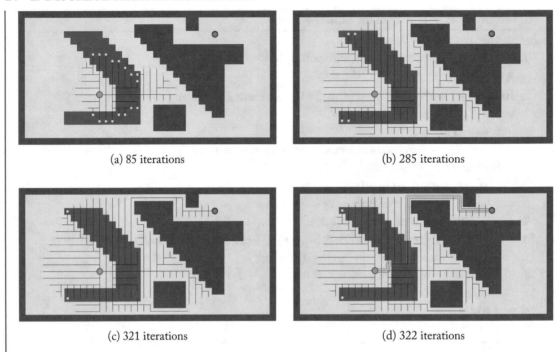

(a) 85 iterations   (b) 285 iterations

(c) 321 iterations   (d) 322 iterations

Figure 2.4: A* expansions while searching for a path from the green sphere to the red sphere. The generated search tree is shown in black and the nodes (leafs) in Q are highlighted in yellow.

If the heuristic function h never overestimates the actual minimal cost of reaching the goal, then an optimal solution will be always found with A*. Intuitively, if $h(n) = 0$ for all n the algorithm reverts to the Dijkstra's expansion, which uses no knowledge from the problem domain. It is like any node in the expansion frontier could get to the goal by an arbitrarily low cost edge connection. For problems related to navigation we know that this is not possible and every node will at least have a cost that depends on its straight line distance to the goal in order to reach the goal at some point. Function h encodes that cost. If no more information from the problem domain is available, A* expands no more nodes than any other algorithm that can compute optimal paths. If information is available to improve heuristic estimates, the performance of A* can improve.

A good well-informed heuristic can make a significant difference on the performance of A*. The straight-line distance heuristic predicts that a node can move along a straight line toward the goal, a prediction that will fail every time there is an obstacle between the current node and the goal, resulting in a large number of nodes being expanded. There are simple strategies to improve this heuristic for our examples. For instance, in our 4-connected grid a *Manhattan distance* provides a better estimate for the obtained paths and can be defined by $x + y$, where x and y are the respective x-distance and y-distance from a node to the goal. In an 8-connected grid, a simple *octile heuristic* can be defined to provide a better estimate by taking into account the fact that paths can only have diagonal sections at 45° from the cardinal directions. This heuristic can be defined as $max(x, y) + (\sqrt{2} - 1)min(x, y)$. These heuristics improve the performance of A* in the presented example environment. An overview of these and a number of additional techniques for encoding better-informed heuristics are discussed in the next section.

2.3   EXTENSIONS AND ADDITIONAL METHODS

Dijkstra and A* are generic graph search algorithms that can run in $\mathcal{O}(n \log n)$ time on a variety of graph types. Given their flexibility and simplicity of implementation these algorithms are widely used for navigation problems. A number of additional extensions and implementation choices are, however, important to be considered.

2.3.1   GRID CONNECTIVITY

The presented examples in grid environments (Figures 2.2 and 2.4) were based on 4-connected grids, i.e., on grids with cells considered to be connected only to the adjacent cells along the horizontal and vertical cardinal directions. The main drawback of using 4-connected grids is that they lead to paths only composed of vertical and horizontal segments. A practical way to improve path quality is to work with 8-connected grids. In this case cells are connected to their diagonal neighbors in addition to their cardinal neighbors. This results in better paths because diagonal moves can also be considered.

Figure 2.5 shows results obtained with 8-connected grids on the same grid environments of the previous examples. The obtained paths are of better quality and they are also shorter in length because one diagonal move is shorter than two equivalent cardinal moves. The used costs were 1.0 for a cardinal move and 1.4 for a diagonal move. These costs are then multiplied by the terrain traversal cost respective to each move.

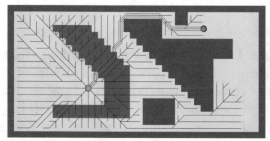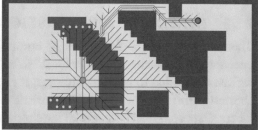

Figure 2.5: Path quality is improved when considering 8-connected grids instead of 4-connected grids; however, imposing a higher number of edges to be processed. The search state of Dijkstra (left) and A* (right) at the solution iteration are shown.

The drawback of using 8-connected grids is that the number of edges to be processed is increased. Even if the number of iterations to reach a solution may not increase, the iterations will on average expand more nodes since there are more neighbors per vertex. The result is an increased number of queue operations. In the presented examples, Dijkstra's algorithm required 33 more iterations to reach a solution path. A* was not trapped by any obstacle and was able to progress toward the goal in fewer iterations than in the 4-connected grid. However, both Dijkstra and A* required more queue operations than in the 4-connected grid. Table 2.1 summarizes the obtained numbers. The table also shows that a higher graph connectivity translates into an increased number of decrease-key operations.

Table 2.1: Performance of algorithms in 4-connected (4-c) and 8-connected (8-c) grids

	4-c Dijkstra	8-c Dijkstra	4-c A*	8-c A*
Iterations until solution:	486	519	322	294
Decrease-key operations:	0	56	16	106
Total queue operations:	494	534	355	378

The numbers in Table 2.1 were obtained for A* using the straight line heuristic. In the 4-connected grid, the number of iterations for A* to reach a solution path decreases from 322 to 267 when using the Manhattan heuristic. In the 8-connected grid, the number of iterations decrease from 294 to 276 when using the octile heuristic. These numbers illustrate the impact obtained by the discussed techniques; however, the obtained results may vary according to the grid size and the complexity of the obstacles in given environments. The overall chosen approach should take into account all these variables in addition to the desired path quality. Finally, one additional approach that is equivalent to improving the number of different orientations a path can have is proposed by Theta* [Daniel et al., 2010], where angle ranges are propagated at each node expansion, reaching solutions that are closer to the true shortest path.

2.3.2   IMPROVING A* HEURISTICS

The use of a heuristic cost makes A* outperform Dijkstra in most of the navigation problems encountered in practice. In order for A* to compute optimal solutions the heuristic cost $h(n)$ for a given node n has to be admissible, i.e., it cannot overestimate its distance to the goal along the shortest path to it. An admissible heuristic will be *consistent* if $|h(n_1) - h(n_2)| \leq dist(n_1, n_2)$, where $dist(n_1, n_2)$ is the length of the shortest path between n_1 and n_2. In principle, a consistent heuristic should lead to better results because it better reflects what a meaningful heuristic function should encode. However, it is common in practice to rely on consistent heuristics which can become poor estimators in several situations, as is the case of the simple heuristics discussed in the previous examples. When the search is trapped behind an obstacle, since the heuristic is consistent, it will have to expand all the trapped nodes in a Dijkstra-like fashion until the search is able to escape the region of poor heuristic values to then be able to move directly toward the goal.

One option is to explore the use of admissible but not consistent heuristics. In this case, when states with poor heuristic estimation are reached, they may have a larger heuristic value than their parents and may possibly escape from the poorly estimated region of the search space before having to expand many other nodes. The drawback of an inconsistent heuristic is that a same node may end up expanded more times than with a consistent heuristic, increasing the number of queue operations and the total number of iterations. Nevertheless, experiments have been shown where inconsistent heuristics can improve the performance of A* and other search algorithms [Felner et al., 2011].

An alternative way to define the prioritization of nodes in Q is to use the weighted function $w_1 g(n) + w_2 h(n)$ instead of simply $g(n) + h(n)$. This introduces two weights that can be used to adjust the behavior of the search. When $w_1 = 1$, we obtain Dijkstra with $w_2 = 0$ and A* with $w_2 = 1$. When $w_1 = 0$ and $w_2 = 1$, we obtain pure heuristic search, a greedy best-first search which can be quick in certain situations; however, most likely not finding the optimal path to the goal. When $w_1 = 1$ and $w_2 > 1$ we obtain the so-called weighted A*, which also gives up optimality to try to find solutions more quickly. Another area of improvement is to efficiently handle tie-breaking. When there are many states with the same costs in Q, a specific order to choose among them can make a significant difference in some cases, such as expanding the ones with highest g-cost first. A number of heuristics and additional techniques can be found in the heuristic search research literature [Goldberg and Harrelson, 2005], including heuristics that rely on detecting properties of the environment (such as dead-ends and gateways) and that involve pre-processing environments in advance [Björnsson and Halldórsson, 2006, Rayner et al., 2011].

2.4   EXTENSIONS AND ADDITIONAL METHODS

One important observation to make is that most of the techniques developed by the heuristic search community target practical improvements to search on grid environments, which have particular representation properties that are explored by the proposed methods. Research on graph algorithms instead mostly address algorithms that run on generic graphs. It is notable that shortest paths can be found in planar graphs with non-negative edge lengths in $\mathcal{O}(n)$ running time [Klein et al., 1994]. The approach relies on multiple priority queues and requires decomposing the graph into small regions and sub-regions. Because each of these regions is small, the priority queues are kept small and thus queue operations are cheap. These and other variations are expected to become important as search algorithms transition from traditional grid environments to navigation mesh environments, where adjacency graphs usually have highly irregular edge costs connecting a much lower number of nodes.

A number of additional extensions and techniques are available to the search algorithms described in this chapter. Extensions for addressing dynamic environments, for planning under time constraints, and for speeding up the search with hierarchical representations are discussed in Part III.

PART II

Geometric Algorithms and Environment Representations

CHAPTER 3

Euclidean Shortest Paths

In the previous chapters we introduced the problem of representing a planar environment such that graph search algorithms can be employed for path planning. Because most applications of virtual worlds will focus on speed of computation, graph representations will most often be of linear size even though the shortest paths computed in them will often not be the shortest path in the corresponding Euclidean plane. The problem of finding Euclidean shortest paths is a classical problem in computational geometry and this chapter reviews the main approaches for solving it.

Let p and q be points in $\mathbb{R}^2$ and let $\mathcal{S}$ denote the set of all polygonal obstacles (in $\mathbb{R}^2$) that are in consideration. The Euclidean Shortest Path (ESP) between p and q is a path of minimum length among all possible paths connecting p to q without crossing any obstacles in $\mathcal{S}$. The ESP is therefore globally optimal, meaning that there is no other collision-free path of shorter length. It is possible that more than one ESP exists, but in such case all of them will have the same minimum possible length.

The ESP problem does not take into account irregular terrain costs and optimality is only determined by the path length, which is measured by the sum of the Euclidean distances between adjacent points in the path. If a ESP is not a straight-line segment connecting p and q, then the path will have to pass by vertices of the obstacles in $\mathcal{S}$. Except for the end-points of the path, the path will be described by a sequence of obstacle vertices.

In the sections below we will denote ESPs with $\pi^*(p, q)$. If a collision-free path exists without any guarantees that the path is a globally minimum one, we will call the path as $\pi(p, q)$. The total number of vertices used to describe all the obstacles in $\mathcal{S}$ is n.

3.1 EUCLIDEAN SHORTEST PATHS IN SIMPLE POLYGONS

The simplest version of the ESP problem is when $\mathcal{S}$ is reduced to a single simple polygon with p and q given inside the polygon. While this situation is usually too restrictive for navigation problems, it serves as an introduction to ESP problems and, as we will see later, computing paths in polygons has become important for optimizing paths in given corridors of a generic environment.

When the region of interest is reduced to a simple polygon, since there are no "holes" in the considered region, all possible paths connecting the two points can be continuously deformed into the globally optimal path π^*. A good concept to use is that of an elastic band that deforms itself to its minimal length state without crossing any edges in $\mathcal{S}$. The result will be $\pi^*(p, q)$. This property eliminates the need for searching among different corridors that could connect p to q, and the efficient *Funnel* algorithm [Chazelle, 1982, Hershberger and Snoeyink, 1994, Lee and Preparata, 1984] can be employed to compute π^* in optimal $\mathcal{O}(n)$ time.

The Funnel algorithm assumes that the polygon is first triangulated, what can also be achieved in linear $\mathcal{O}(n)$ time [Amato et al., 2000, Chazelle, 1991]. In practice, however, these optimal triangulation algorithms are difficult to implement and they do not seek for a triangulation of good quality. For these reasons, other triangulation methods such as the Delaunay Triangulation (presented later in Section 4.1.3) are often preferred even if the computational time increases to $\mathcal{O}(n \log n)$. Figure 3.1 exemplifies a triangulated simple polygon and the ESP between two points inside it.

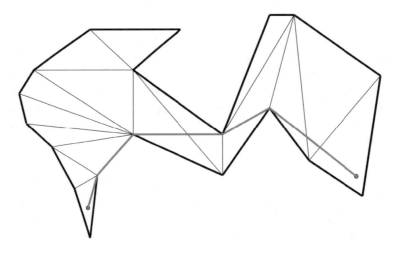

Figure 3.1: The Euclidean Shortest Path between two points inside a triangulated simple polygon.

The algorithm performs a sequential expansion of the triangulation edges while maintaining a funnel structure (see Figure 3.2) representing the visibility regions of the vertices processed so far. The triangulation gives the edge expansion order and for each edge expanded one new vertex is processed. Let p be the starting point of the shortest path being computed and u and v be the polygon vertices at the extreme edge of the funnel, that is, the *door* of the funnel, illustrated as edge e in Figure 3.2. Paths $\pi^*(p, v)$ and $\pi^*(p, u)$ may travel together for a while, and at some vertex a they diverge and are concave until they reach the door vertices u and v. The funnel is the region delimited by segment uv and the concave chains $\pi^*(a, v)$ and $\pi^*(a, u)$, and a is its apex. The vertices of the funnel are stored in a double-ended queue Q for efficient processing. Each side of Q processes vertices being added to one of the concave chains composing the funnel.

Figure 3.2 illustrates the insertion process of a new vertex w. Points from the v end of Q are popped until b is reached, which is the vertex that will maintain a concave chain to w. If the apex of the funnel is popped during the process, it becomes part of the path so far and a new apex that maintains the funnel structure is determined. The process advances until reaching the destination point q. When q is reached, q is connected to either the apex or, if the apex is not visible, to the visible vertex in the boundary of the funnel that will complete the shortest path.

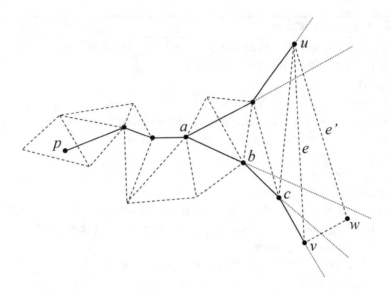

Figure 3.2: The Funnel algorithm uses a triangulation to process vertices in a correct order. The diagram illustrates the insertion process of a new vertex w.

Algorithm 5 provides a more detailed description of the Funnel algorithm. The funnel door e is advanced along the triangulation T of the input polygon until the destination point q is reached. The notation assumes T to be in a horizontal layout, as in Figure 3.2, and $e.topv$ and $e.botv$ refer to the top and bottom door vertices, respectively. The double-ended queue Q contains the vertices of the funnel being processed. Top vertices are pushed to the left side of Q and bottom vertices are pushed to the right side of Q. After initialization, the algorithm enters the the main loop where the next door edge e' is determined, q is tested to be reached and, if not reached, the funnel is adjusted to contain the new vertex reached by e'. The adjustment (lines 10–22) is performed according to two cases. In the first case, $e'.topv$ is added to the top side of the funnel (pushed to the left side of Q), but before that, the vertices on the left side of the funnel are popped as needed in order to maintain a concave left chain. This process relies on a CW primitive, which tests if the chain $(Q.left(1), Q.left(0), e'.topv)$ has clockwise orientation, where $Q.left(0)$ and $Q.left(1)$ return the top-most and second top-most vertices from the left end of Q, respectively. The second case performs the equivalent operations but symmetrically in order to add $e'.botv$ to the bottom side of the funnel (right side of Q), and where the used CCW primitive denotes a counter-clockwise test.

If the triangles in T are exhausted and q is not reached, Algorithm 5 returns a null path. This can happen if the adjacency graph of T has several branches and the algorithm traverses the wrong one. To avoid such cases, the input triangulation should be pre-processed in order to determine a single corridor of triangles to be processed.

Algorithm 5 - Funnel algorithm

Input: simple polygon triangulation T and query points p and q in T.
Output: $\pi^*(p, q)$, or null path if q not reachable.

1: **Funnel**(p, q)
2: Initialize double-ended queue Q with p;
3: Initialize path π^* with p, and mark p to be the initial funnel apex;
4: $e \leftarrow$ first edge of T in front of p; // e is the current funnel door
5: $Q.push\_left(e.topv)$;
6: $Q.push\_right(e.botv)$;
7: **while** (e has an adjacent non-visited triangle in T) **do**
8:    $e' \leftarrow$ next non-visited interior edge of T sharing a triangle with e;
9:    **if** (q is in triangle shared by e and e') { finalize path to p; **return** π^*; }
10:   **if** ($e'.botv = e.botv$) **then**
11:      **while** (CW$(Q.left(1), Q.left(0), e'.topv)$) **do**
12:         $v = Q.pop\_left()$;
13:         **if** (v is apex) { Add v to π^*, and set $Q.left(0)$ as new apex; }
14:      **end while**
15:      $Q.push\_left(e'.topv)$;
16:   **else**
17:      **while** (CCW$(Q.right(1), Q.right(0), e'.botv)$) **do**
18:         $v = Q.pop\_right()$;
19:         **if** (v is apex) { Add v to π^*, and set $Q.right(0)$ as new apex; }
20:      **end while**
21:      $Q.push\_right(e'.botv)$;
22:   **end if**
23:   $e \leftarrow e'$;
24: **end while**
25: **return** null path;

Computing shortest paths in simple polygons is usually not enough for navigation problems because environments based on a single simple polygon are too restrictive. However, the Funnel algorithm has become particularly useful as a post-processing technique after a corridor containing a solution path is determined. First, a graph search algorithm is applied on the adjacency graph of a generic polygonal cell decomposition. The solution determines the polygonal cells that contain the solution path. These cells can be triangulated and their union will form a triangulated simple polygon describing the corridor containing the solution path. The Funnel algorithm can then be applied in order to determine the shortest path in the corridor. While this shortest path may not be the globally shortest one with respect to the whole environment, it will be a locally shortest one, the shortest with respect to the corridor.

3.2   VISIBILITY GRAPHS

While ESPs in simple polygons can be efficiently computed, the generic case is harder because of the many corridors that may exist in a given environment. Probably the most popular and well-known approach for computing ESPs in generic planar polygonal environments is to build and search the *visibility graph* [de Berg et al., 2008, Lozano-Pérez and Wesley, 1979, Nilsson, 1969] of the obstacles in S.

The visibility graph is the graph composed of all segments connecting vertices that are visible to each other in S. While simple implementations will often rely on $\mathcal{O}(n^3)$ or $\mathcal{O}(n^2 \log n)$ algorithms, visibility graphs can be computed in $\mathcal{O}(n^2)$ time [de Berg et al., 2008] and in the generic case this time cannot be reduced because the graph itself has $\mathcal{O}(n^2)$ edges. Graph search algorithms can be then applied to the visibility graph for finding ESPs. Because an ESP is composed of edges connecting vertices in S, and all possible valid edges connecting vertices are in the visibility graph, all possible paths are represented in the graph and the lowest-cost path in the graph will be the shortest path in the Euclidean plane. Both Dijkstra and A* algorithms (Chapter 2) can be applied.

In order to use the visibility graph to compute $\pi^*(p, q)$, points p and q have first to be connected to the graph by augmenting the graph with edges connecting all visible vertices to p and q; see Figure 3.3. The result is the same as if points p and q were points in S. After the graph is augmented, the lowest-cost path in the graph between p and q will be $\pi^*(p, q)$.

A straightforward optimization for reducing the number of edges in the visibility graph is to not include edges that lead to *non-convex* vertices. These edges can be safely discarded because shortest paths will always pass by convex vertices. This optimization can significantly reduce the number of edges, as shown in Figure 3.4, however the total number of edges remains $\mathcal{O}(n^2)$. Visibility graphs represent the most straightforward way to compute ESPs and extensions for handling arbitrary clearance have also been developed [Chew, 1985, Liu and Arimoto, 1995, Wein et al., 2007].

Algorithm 6 illustrates a possible strategy for using a visibility graph G for multiple path queries. First, G is augmented with edges to take into account the query points, then the solution path is computed, and finally G is restored to its initial state before returning the solution path. This ensures that G can be re-used for multiple queries. Note that the visibility graph may have disconnected components and if the query points are not connected to a same component a path will not be found.

The main difficulty with visibility graphs is that path search in a graph that may contain a quadratic number of edges is in general too expensive for interactive applications. This is a consequence of the combinatorial complexity of the ESP problem. The approach taken by the visibility

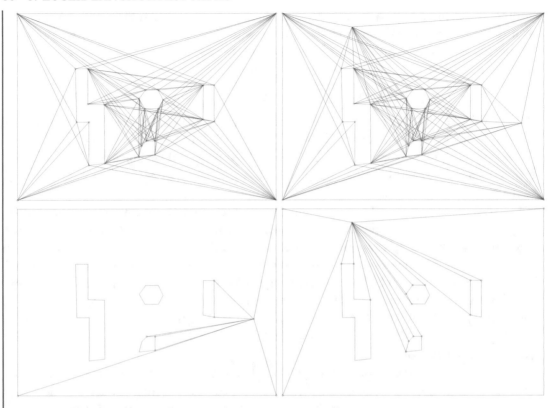

Figure 3.3: The Euclidean shortest path between p and q can be found by searching the visibility graph of S (top-left) augmented by the edges connecting all visible vertices to p and q (top-right). The diagrams on the bottom show the added edges connecting the visible vertices in the graph to query points p and q.

Algorithm 6 - Path Search on Visibility Graph

Input: visibility graph G and query points p and q.

Output: $\pi^*(p,q)$, or null path if no path exists.

1: **PathFromVG**(p,q)
2: Add edges to all convex vertices in G that are visible to p;
3: Add edges to all convex vertices in G that are visible to q;
4: $\pi^* = \text{AStar}(p,q)$;
5: Remove from G all the added edges;
6: **return** π^*;

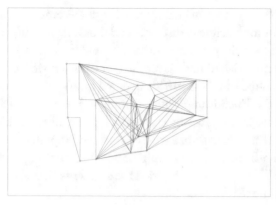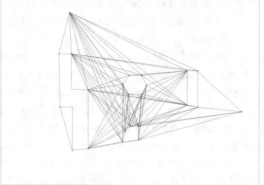

Figure 3.4: A popular optimization to reduce the number of edges in the visibility graph is to discard edges that lead to *non-convex* vertices, both when processing the environment (left) and when augmenting the graph with the source and destination points (right). In this example the discarded edges were the ones connecting to the environment boundary, and to the two concave vertices in the left-most obstacle.

graph is to capture all possible segments that ESPs can have and let graph search algorithms determine the shortest path between two of its nodes. When Dijkstra or A* is applied to a visibility graph the running time will be $\mathcal{O}(n^2 \log n)$. Another difficulty is to update the visibility graph when there are changes to the obstacle set. Because a single vertex may be connected to all others, an update may involve global visibility computations on the entire obstacle set. The visibility graph is not a structure with local connectivity properties and repairing operations will most often not be local.

3.3   THE SHORTEST PATH TREE

Visibility graphs require potentially expensive graph searches in order for paths to be retrieved from them. One structure that is useful in situations where multiple path queries are needed for a same source point is the Shortest Path Tree (SPT).

The SPT is a structure specific to a given source point p. The SPT is the tree formed by all ESPs from p to each reachable vertex in the environment. The SPT is a subgraph of the visibility graph and is usually computed from the visibility graph by running an exhaustive Dijkstra expansion starting from the source node representing p, until there are no more nodes reachable from p. Given polygonal obstacles S, the visibility graph for S is first computed and augmented with the edges connecting p to all visible vertices in S. Let s denote the source node representing p in the visibility graph. Dijkstra's expansion can be then launched starting from s.

An implementation of the Dijktra's expansion is illustrated in Algorithm 7. It uses the same notation as in Algorithm 1 (Chapter 2). The expansion will visit all reachable nodes in the graph by expanding a frontier, stored in priority queue Q, of approximately equal cost-to-come from s. Here the cost-to-come to a node n only accounts for Euclidean distances and thus it represents the length of $\pi^*(s, n)$. This length is often referred to as the *geodesic distance* to n, and in Algorithm 7 it is retrieved with function $g(n)$. Every time a non-visited node is reached, or a node is visited by a shorter path, the SPT parent node of n is updated to v, and $g(n)$ is updated to $g(v) + d(v, n)$, where v is the parent node of n and $d(v, n)$ is the Euclidean distance from v to n.

The expansion continues until there are no more nodes in Q, at which point all reachable nodes from s have been processed. At the end of the algorithm, each processed node n will have a pointer to its parent in the SPT. The connection from n to its parent represents the first segment along the ESP from n to s. To visualize the SPT it is enough to draw the segments connecting each processed node to its SPT parent; see Figure 3.5 for examples.

Algorithm 7 - SPT Computation by Dijkstra's Expansion

Input: visibility graph and source node s.

Output: shortest path tree rooted s.

1: **SPT**(s)
2: Initialize Q with $(s, 0)$, set $g(s)$ to be 0, and mark s as visited;
3: **while** (Q not empty) **do**
4:    $v \leftarrow Q.remove()$;
5:    **for each** (neighbors n of v) **do**
6:        **if** (n not visited **or** $g(n) > g(v) + d(v, n)$) **then**
7:            Set the SPT parent of n to be v;
8:            Set $g(n)$ to be $g(v) + d(v, n)$;
9:            **if** (n visited) $Q.decrease(n, g(n))$; **else** $Q.insert(n, g(n))$;
10:           Mark n as visited, if not already visited;
11:       **end if**
12:    **end for**
13: **end while**

The SPT is specific to a given source point p but it has the advantage that it does not require a graph search to determine paths. Since it is a tree, ESPs to vertices can be computed in time proportional to the number of vertices in each ESP, by tracing back the SPT parents from any given vertex until reaching the root node of the SPT.

More interestingly, the SPT can be efficiently used to compute ESPs from the source point p to amy generic target point q in the plane. In this case, each geodesic distance $g(n)$ computed during the construction of the SPT remains stored in each vertex of the SPT. Given target point q, the set V of vertices in the SPT that are visible to q is first computed. Then, $v \in V$ is determined such that $g(v) + d(v, q)$ is minimum. The ESP from p to q, or $\pi^*(p, q)$, is obtained by appending q to the SPT branch from s to v. The overall process is summarized in Algorithm 8. The +

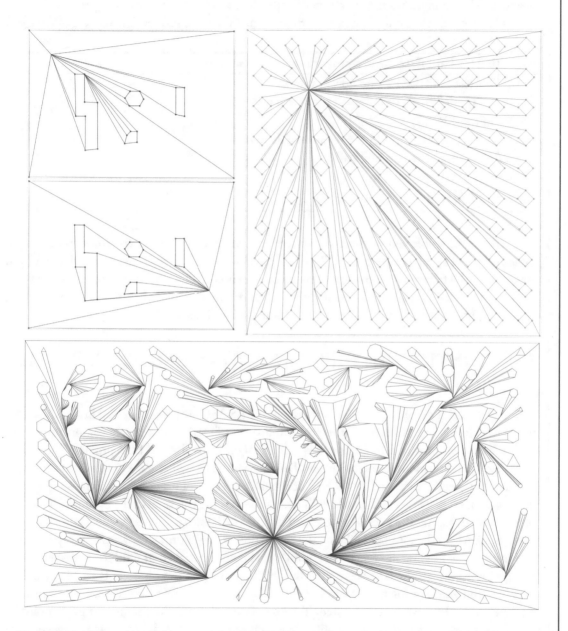

Figure 3.5: Examples of shortest path trees (SPTs) rooted at arbitrary points in different environments. The SPT contains all Euclidean shortest paths from its source point to each reachable vertex in the environment.

operator in line 5 indicates the operation of appending a point to the path being returned. As in the visibility graph case, only convex vertices are required to be considered as vertices of the SPT.

Algorithm 8 - Path Extraction from SPT

Input: Shortest Path Tree T and query point q.
Output: $\pi^*(root(T), q)$, or null path if no path exists.

1: **PathFromSPT**(q)
2: $V \leftarrow$ all convex vertices in T that are visible to q;
3: **if** (V is empty) **return** null path;
4: $v \leftarrow \min_{v \in V} g(v) + d(v, q)$;
5: **return** $branch(root(T), v) + q$;

The set of visible vertices from a given point q can be computed in $O(n \log n)$ time using a rotational plane sweep [de Berg et al., 2008]. It can also be computed from a triangulation of the environment by traversing all triangles with visible portions from q by adjacency, outwards from the triangle containing q. This is usually efficient in practice because large non-visible areas can be pruned, however the approach has $\mathcal{O}(n^2)$ computation time.

3.4   CONTINUOUS DIJKSTRA AND THE SHORTEST PATH MAP

Although visibility graphs may have $\mathcal{O}(n^2)$ edges, the ESP problem can be solved in sub-quadratic time by exploring the planarity of the problem instead of relying in a purely combinatorial approach. Mitchell provided the first sub-quadratic algorithms for the ESP problem [Mitchell, 1991, 1993] and introduced the *continuous Dijkstra* paradigm [Mitchell, 1993], which simulates the propagation of a continuous wavefront of equal length to a source point, without using the visibility graph. Hershberger and Suri [1997] later improved the run time to the optimal $\mathcal{O}(n \log n)$ time, while increasing the needed space from $\mathcal{O}(n)$ to $\mathcal{O}(n \log n)$.

Similarly to the discrete Dijkstra, the simulation of the continuous wavefront expansion is performed by processing discrete events, until a goal point is reached or until the whole environment is processed. In the latter case the result of the process is the Shortest Path Map (SPM), a spatial subdivision of the free region in $\mathcal{O}(n)$ cells such that the ESP to all reachable points in the Euclidean plane can be represented. The boundaries of the SPM delimit, for every point in the plane, to which vertex the point should be connected first in order to construct its shortest path to the source vertex. For this to be possible, boundaries of the SPM may be hyperbolic arcs generated by self-collisions of the expansion front. The result is a remarkable decomposition able to efficiently encode all possible ESPs, and despite the fact that shortest paths are a well-known concept, the SPM is usually unknown to those outside the computational geometry field.

Figure 3.6 illustrates the continuous Dijkstra paradigm, its simulation of the continuous wavefront propagation, and the resulting SPM. The circular wavefront emanates from the source vertex p and is updated by events caused by vertices being reached or by the front colliding with itself. Figure 3.6 illustrates the initial steps of the process. The first vertex encountered by the wavefront is vertex a, which breaks the wavefront into a circular arc that starts and ends at the edges adjacent to a. The circular arc continues to grow, it subdivides when an obstacle edge is reached, and it slides along the contact edges until the next event. The second vertex encountered is vertex b, which is adjacent to one visible and one non-visible polygon edge. Because of that b becomes a new front generator. This means that b becomes the center of a new circular arc expanding toward a previously unseen region, and the total expansion frontier is now composed of two expanding circular arcs. New generators are also encountered for the next two vertices c and d. Each new generator becomes the center of a new circular arc that becomes part of the overall expansion frontier. When the front collides with itself the collision is between a pair of expanding circular arcs, what generates an hyperbolic arc that is equidistant to the two generators of the colliding fronts. The hyperbolic arcs will become boundaries of the SPM, as it can be seen in the last two images. The other boundaries of the SPM are straight-line segments delimiting the visibility regions of generators. The last image in the picture shows an ESP reconstructed from a given target point q back to the source.

The continuous Dijkstra approach illustrated in Figure 3.6 is intuitive but its efficient implementation involves several complex geometric operations [Mitchell, 1993] that make it difficult to be used in practice. New GPU techniques recently developed [Camporesi and Kallmann, 2014] lead to a practical alternative for computing SPMs in applications. Because SPMs encode shortest paths from a source point to all points in the plane, they are in particular interesting when several paths for a same source point are needed.

The SPM examples presented in this chapter were all computed using GPU shaders [Camporesi and Kallmann, 2014]. The method first computes the SPT of the vertices. The SPT contains at each vertex v the geodesic distance $g(v)$ to the source vertex s. Then, clipped cones simulating an expansion front are placed at each generator vertex v, but below v, at the height given by $g(v)$. Cones are clipped so that they only represent the visible regions from v. The process is illustrated in Figure 3.7. After the clipped cones are correctly placed at the vertices (left image), the cones are rendered from an orthographic vertical camera placed above the obstacle plane (center image). The result in the frame buffer will encode the SPM with respect to the source point (right image).

The cones are rendered with flat colors, each representing the id of its region and generator vertex. Each pixel in the frame buffer therefore contains the id of each generator and locating the region containing a given query point becomes a constant operation. Given a located query point, it is then connected to the generator point of the region containing it, then progressively connected to the parent generators until reaching the source node. The traversed sequence of points is the Euclidean shortest path to the source vertex. With this GPU rendering approach,

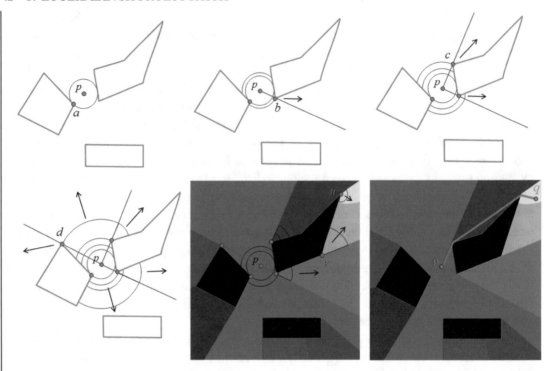

Figure 3.6:   In left-right, top-bottom order: the shortest path map can be computed by simulating the continuous propagation of a circular wavefront emanating from source vertex p. The wavefront is updated by events caused by vertices being reached or by the front colliding with itself, in which case a hyperbolic frontier is generated. An ESP for a given point q can be computed by connecting q to the generator of the region containing it, and then tracing back parent generators until reaching the source point.

hyperbolic frontiers are naturally obtained with Z-Buffer operations when rendering intersecting cones.

Figure 3.8 shows SPMs computed for different environments. At the time of writing the described GPU method had been updated in several ways, with dedicated fragment shaders for efficient cone rendering, and with *compute shaders* to compute both the SPT and the SPM in a single pass.

The SPM can be seen as a generalization of the SPT to every reachable point in the continuous plane. Given a target point q, once the cell of the SPM containing q is localized, the ESP from q to the source point is retrieved in time proportional to the number of vertices in the ESP. The retrieval process is equivalent to the one for the SPT structure, but without requiring the connection of q to all visible vertices. In the SPM case it is sufficient to localize the query

Figure 3.7: Computing the SPM with GPU rendering. Clipped cones are placed below vertices (left) and then the scene is rendered from an orthographic vertical camera placed above the obstacle plane (center). The result in the frame buffer encodes the SPM (right). The shown images were rendered with Phong shading. The final SPM will use flat colors encoding region ids.

point because each cell is associated with its one generator vertex, and each generator vertex keeps a direct pointer to its parent generator vertex. Algorithm 9 summarizes the main steps. The + operator in line 4 is used to indicate that q is appended to the path being returned.

Algorithm 9 - Path Extraction from SPM

Input: Shortest Path Map M and query point q.

Output: $\pi^*(root(M), q)$, or null path if no path exists.

1: **PathFromSPM**(q)
2: **if** (q not in a region of M) **return** null path;
3: $v \leftarrow$ generator vertex of region containing q;
4: **return** $branch(root(M), v) + q$;

3.5 EXTENSIONS AND ADDITIONAL METHODS

Several extended versions of the Euclidean shortest path problem exist. Two particular extensions that are of particular relevance to navigation are to compute globally optimal paths in terrains with irregular traversal costs, and to compute ESPs with clearance from obstacles.

The computation of globally optimal paths across regions of varied traversal costs entails solving the Weighted Region Problem (WRP) [Mitchell and Papadimitriou, 1991]. Regions are described by polygonal boundaries and each region has a traversal cost weight that is to be multiplied by the length of a path traversing the region. The total cost of a path will take into account the weights of all traversal regions. The optimal path can be computed with an extension of the continuous Dijkstra algorithm, where the wavefront will refract each time it crosses the boundary

Figure 3.8: Examples of shortest path maps (SPMs) for different environments. Each cell is represented by a different color. On the two top environments the location of the source point is denoted with a cross. On the bottom environment the source point is p and the query point is q. The ESP from p to q is shown and the location of each generator vertex along the ESP is indicated by a black dot. Each generator stores the id of its parent generator, allowing quick path reconstruction in time proportional to the number of vertices in the path.

of two regions with different costs, following Snell's rule of refraction. Practical implementations of such optimal algorithm based on polygonal regions are difficult to achieve and practical approaches are typically based on grid approximations, as seen in Chapter 1.

One approach to include clearance from obstacles in the computation of an ESP is to first inflate the obstacles by the clearance value, and then compute the ESP in the inflated environment. The result will be a path tangent to the inflated obstacles and therefore with the desired clearance to the original obstacles. This approach requires pre-processing the environment specifically for one given clearance value. One particular structure that was designed for computing ESPs for arbitrary clearance values is the Visibility-Voronoi Complex [Wein et al., 2007], which builds a visibility graph that can be quickly updated in order to take into account any clearance value. This method has been implemented and is a suitable alternative for computing ESPs with clearance; however, with computation and query times of $\mathcal{O}(n^2 \log n)$.

The next chapter reviews geometric structures that offer efficient path computation and that are suitable for handling clearance; however, without addressing the computation of globally shortest paths.

CHAPTER 4

Navigation Meshes and Geometric Structures with Clearance

A navigation mesh is a polygonal decomposition of the navigable space in a virtual world, and which is designed specifically to delimit navigable regions and to efficiently support the computation of navigation queries. By delimiting the free navigable regions, the navigation mesh can support path planning and also provide important spatial information that agents can use during collision avoidance and behavior execution. As seen in Chapter 1, this is a key advantage that navigation meshes offer over graph-based representations such as waypoint graphs or roadmaps. Such advantages were the motivation behind early work on designing navigation meshes [Snook, 2000, Tozour, 2002]. However, while the term navigation mesh has become well accepted and widely used, no formal definition or guaranteed properties are attached to it. There is a tendency on using triangulations, of any kind, but this also is not a requirement. Most of the structures seen in the previous chapters could be used as the underlying representation scheme.

Ideally a navigation mesh would support the computation of paths with any desired properties; however, the supported properties are commonly limited to the ones that can be computed efficiently. For instance, as seen in the previous chapter, no linear-size environment decomposition can efficiently support the computation of Euclidean shortest paths for arbitrary pairs of query points. For this reason, in practical virtual world applications where speed of computation is most important, global optimality of paths is often not supported. Instead, a number of different approaches for navigation meshes have been developed with focus on different properties. One particular property that is important is clearance from obstacles.

In the next sections we first review approaches suitable for geometrically encoding clearance, and later review additional topics and properties that are also important to be addressed by navigation meshes. This chapter is presented as an overview of the several related topics, with focus on methods and approaches with known properties. In-depth analysis and detailed algorithms are left to the cited references.

4.1 GEOMETRIC STRUCTURES SUPPORTING CLEARANCE

In many cases, path clearance from obstacles is considered more important than ensuring that globally shortest paths are computed. There are different approaches that can be used to design representations that can encode clearance information. These structures depart from the goal of finding globally shortest paths, and instead provide structures of linear $\mathcal{O}(n)$ size supporting fast path computation with arbitrary clearance guarantees, where n is the total number of vertices describing all polygonal obstacles in consideration.

4.1.1 THE VORONOI DIAGRAM

The most classical spatial partitioning structure that encodes clearance is the Voronoi diagram [de Berg et al., 2008, Preparata and Shamos, 1985]. The Voronoi diagram is most well known for a set of seed points in the plane. In this case, the Voronoi diagram partitions the plane in cells such that for each seed there will be a corresponding cell consisting of all points closer to that seed than to any other. The boundaries between the cells will be segments that are equidistant to two seeds. The set of boundaries therefore represents a graph of maximum clearance from the seed points.

The Voronoi diagram can be generalized to a set of seed segments, and in this case the Voronoi cells will delimit the regions of the plane that are closest to each segment. Consider now that the segments are the edges of the input obstacles in a given set $\mathcal{S}$. The edges of the generalized Voronoi diagram of $\mathcal{S}$ will be the medial axis of $\mathcal{S}$, which is a graph that completely captures all paths of maximum clearance in $\mathcal{S}$; see Figure 4.1 for an example.

Since the Voronoi diagram is a structure computed for seed points, its generalized version leading to the medial axis is the version that has a direct application to path planning with clearance.

4.1.2 THE MEDIAL AXIS

Path planning using the medial axis graph has been extensively explored. The medial axis is a structure of $\mathcal{O}(n)$ size and search methods can be directly applied to it, therefore efficiently determining shortest paths in the graph after connecting generic source and destination points to their closest points in the graph. The medial axis does not help with finding ESPs but it naturally allows the integration of clearance constraints. Many methods have been proposed for computing paths with clearance by searching the medial axis graph of an environment [Bhattacharya and Gavrilova, 2008, Geraerts, 2010]. The medial axis can be computed from the Voronoi diagram of the environment in time $\mathcal{O}(n \log n)$, and methods based on hardware acceleration have also been developed [Hoff III et al., 2000].

Figure 4.1 illustrates the medial axis of a given environment. Each edge can be associated with its closest elements in the environment and as well with the distance to the closest elements, therefore the clearance that each edge of the graph accommodates can be readily available. Edges that have an obstacle vertex as closest element are parabolic arcs, otherwise the edges are straight line edges. In practice not all edges of the medial graph are useful for path planning and edges that connect to vertices of the environment can be discarded.

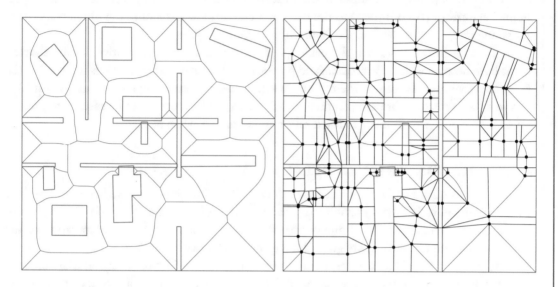

Figure 4.1: The medial axis (left) represents the points of maximum clearance in the environment, and its edges can be decomposed such that each edge is associated with its closest pair of obstacle elements (right). The medial axis is composed of line segments and parabolic arcs.

Given a medial axis graph, the computation of paths with clearance r from obstacles can be reduced to two steps: first, the medial axis graph is searched with the additional constraint that each graph edge in the solution path has to accommodate clearance r, then, if a solution is found, a locally optimal path is determined in the solution corridor. The result is a short and smooth path suitable for navigation, but without global optimality guarantees.

There is a clear distinction between locally and globally shortest paths. Consider a path $\pi(p, q)$ determined after computing the shortest path in the medial axis of S. Path π has maximum clearance and therefore it can become shorter with continuous deformations following the elastic band principle that allows deformations to the path, without allowing it to pass over an obstacle edge, keeping it with clearance r, until it reaches its state of shorter possible length. At this final stage the obtained path will be the shortest path of clearance r in its corridor (or channel), and thus a *locally* shortest path of clearance r, here denoted as $\pi_r^l(p, q)$. A path π^l may or not be the globally shortest one π^*, and as we have seen in the previous chapter, it is not straightforward to determine if π^l is the globally shortest one without the use of appropriate methods and data structures.

Given that the medial axis is computed together with its closest edges and vertices in the environment, the free space of solution corridors is fully represented and decomposed in convex cells; see Figure 4.1-right. A solution corridor can thus be efficiently processed in order to obtain a locally optimal path of clearance r in the corridor, for example by applying an adaptation of the Funnel algorithm (Algorithm 5) to the decomposed corridor. Extending the Funnel algorithm for taking into account clearance is discussed in Section 4.1.4.

Because in real-time virtual worlds speed of computation is imperative, locally shortest paths have been considered acceptable and practically all the navigation methods reported in the literature have been limited to them. One specific benefit of explicitly representing the medial axis in a data structure for path planning is that locally shortest paths can be easily interpolated toward the medial axis in order to reach maximum clearance when needed.

4.1.3   THE CONSTRAINED DELAUNAY TRIANGULATION

Triangulations offer a natural approach for cell decomposition and they have been employed for path planning in varied ways, including to assist with the computation of globally shortest paths. For instance, a triangulated environment has been used to assist with the computation of the relevant subgraph of the visibility graph for a given ESP query [Kapoor et al., 1997]. However, for efficiency and simplicity reasons, methods used in practice will mostly rely on triangulations as a way to decompose the free space to then search for paths on the adjacency graph of the triangulation, as outlined in Chapter 1.

Triangulations are often defined for a given set of input points and not polygons. In order to make sure that the edges of polygonal obstacles will appear as edges in a triangulation of the vertices of the obstacles, the triangulation needs to take into account edge constraints. The most popular type of triangulation with edge constraints is the Constrained Delaunay Triangulation (CDT).

A CDT can be defined as follows. Triangulation T will be a CDT of polygonal obstacles S if: (1) it enforces obstacle constraints, i.e., all segments of S are also edges in T; and (2) it respects the *Delaunay criterion* as much as possible, i.e., the circumcircle of every triangle t of T contains no vertex in its interior which is visible from all three vertices of t. Several textbooks cover the basic algorithms and many software tools are available [Hjelle and Dæhlen, 2006, The CGAL Project, 2014, Tripath Toolkit, 2010].

The CDT is a triangulation that has $\mathcal{O}(n)$ cells and therefore discrete search algorithms can compute channels (or corridors) containing locally shortest solutions in optimal times. Since channels are already triangulated, they are ready to be processed by the Funnel algorithm (Chapter 3). Figure 4.2 illustrates the CDT of a given environment.

One important property of the Delaunay triangulation is that it is the dual graph of the Voronoi diagram. Computing one or the other involves similar work and efficient $\mathcal{O}(n \log n)$ algorithms are available. By being the dual of the Voronoi, CDTs carry an important property that is related to clearance. For every vertex v in a CDT, v will be connected by a CDT edge

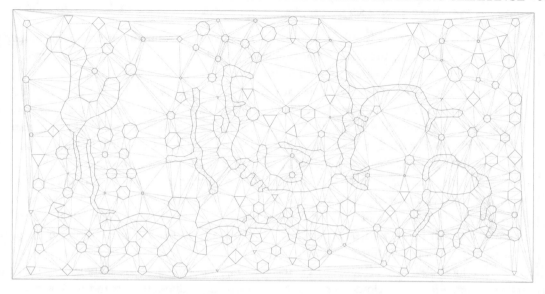

Figure 4.2: The Constrained Delaunay Triangulation provides an $\mathcal{O}(n)$ conformal cell decomposition of any polygonal planar environment given as input. Adjacent triangle cells on the free region of the environment form an adjacency graph that is suitable for path planning.

to its closest visible vertex. This ensures that triangles will connect vertices that are close to each other and is also related to the fact that CDT triangles maximize the minimum interior angle in all triangles. These properties lead to triangles with in general good shape; however, because long obstacle edges can exist, triangles can still be very long and skinny, as it can be observed in the triangles connecting to the outer boundary of the environment in Figure 4.2.

Despite the interesting properties, there is no always-efficient constant time operation that can be used to extract clearance at any location in a CDT. It is possible to check for clearance by visiting adjacent triangles of a given location until reaching the closest obstacle; however, such a test may need to traverse $\mathcal{O}(n)$ elements and it would be expensive to execute it at every expansion step of a path search algorithm.

Fortunately, a recent approach shows that with refinement operations applied to a CDT it is possible to efficiently represent local clearance information in order to efficiently compute correct paths with arbitrary clearance. The refined triangulation is called a Local Clearance Triangulation, and is discussed next.

4.1.4   LOCAL CLEARANCE TRIANGULATIONS

Although $CDT(S)$ is already able to well represent a given environment, an additional property achieved by refinement, the *local clearance property*, has been proposed to support correct and efficient clearance determination for path planning in CDTs [Kallmann, 2014].

Let $T = CDT(S)$ and π be a free path in T between points p and q. Path π is considered free if it does not cross any constrained edge of T. A free path may cross several triangles sharing unconstrained edges and the union of all traversed triangles is called a *channel*. Let t be a triangle in the channel of π such that t is not the first or last triangle in the channel. In this case π will always traverse t by crossing two edges of t. Let a, b, c be the vertices of t and consider that π crosses t by first crossing edge ab and then bc. This particular traversal of t is denoted by τ_{abc}, where ab is the entrance edge and bc is the exit edge. The shared vertex b is called the traversal corner, and the traversal sector is defined as the circle sector between the entrance and exit edges, and of radius $min\{dist(a,b), dist(b,c)\}$, where $dist$ denotes the Euclidean distance. Edge ac is called the interior edge of the traversal.

The local clearance $cl(a,b,c)$ of a traversal τ_{abc} is defined as the distance between the traversal corner b and the closest vertex or constrained edge intersecting its traversal sector.

Because of the Delaunay criterion, a and c are the only vertices in the sector, and thus $cl(a,b,c) \leq min\{dist(a,b), dist(b,c)\}$. In case $cl(a,b,c)$ is determined by a constrained edge s crossing the traversal sector, as illustrated in Figure 4.3, then $cl(a,b,c) = dist(b,s)$ and s is the closest constraint to the traversal. If edge ac is constrained, then ac is the closest constraint and $cl(a,b,c) = dist(b,ac)$. If the traversal sector is not crossed by a constrained edge then $cl(a,b,c) = min\{dist(a,b), dist(b,c)\}$.

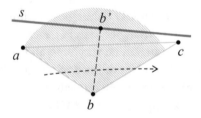

Figure 4.3: The triangle traversal with entrance edge ab and exit edge bc is denoted as τ_{abc}. Segment s is the closest constraint crossing the sector of τ_{abc}, thus $cl(a,b,c) = dist(b,s) = dist(b,b')$, where b' is the orthogonal projection of b on s.

The difficulty is that when a triangle is traversed it is not possible to know how the next traversals will take place: if the path will continue in the direction of a possibly long edge (and possibly encounter a narrower space ahead) or if the path will *rotate around* the traversal corner. Each case would require a different clearance value to be considered. For example, Figure 4.4-left shows an example with long CDT triangles where their clearance values are not enough to capture the clearance along the direction of their longest edges. LCT refinements are then proposed to fix this problem by detecting the undesired narrow passages and breaking them down into sub-traversals until a single clearance value per traversal can handle all possible narrow passages.

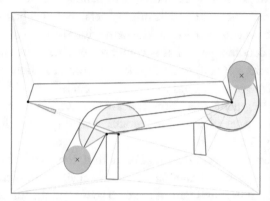 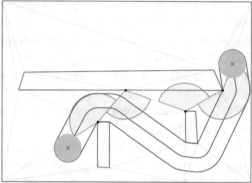

Figure 4.4: The left triangulation is a CDT showing an illegal path that however satisfies all its local clearance tests per traversed triangle. The traversal sectors are highlighted and they all have enough clearance. This example shows that local clearance tests per traversal are not enough in CDTs. However, once the existing disturbances are solved and the corresponding LCT is computed (triangulation on the right), local clearance tests become sufficient.

The vertices that cause undesired narrow passages are called disturbances. A vertex v is detected to be a disturbance if its distance to the constraint of a traversal is shorter than the traversal clearance value. Disturbances can be then detected in the triangulation and the affected traversals can be subdivided with refinement operations until no more disturbances exist. For details on the procedure we refer the reader to Kallmann [2014] that fully describes the method. A traversal is said to have local clearance if it does not have disturbances. A local clearance triangulation is achieved when all of its traversals have local clearance.

Given a desired clearance radius r, the achieved local clearance property guarantees that a simple local clearance test per triangle traversal is enough for determining if a path π can safely traverse a channel with clearance r from constraints. Path π will have enough clearance if $2r < cl(a, b, c)$ for all traversals τ_{abc} of its channel. Furthermore, the clearance value of all traversals in the environment are explicitly stored in the triangulation, reducing clearance tests to a simple floating point comparison. Figure 4.4 presents an example where local clearance tests are not enough to produce correct results in a CDT, while correct results are obtained in the corresponding LCT.

Once a LCT of the environment is available, a graph search can be performed over the adjacency graph of the triangulation in order to obtain a channel of arbitrary clearance r connecting two input points p and q. During graph search in the triangulation adjacency graph, a search expansion is only accepted if the clearance of the traversal being expanded is greater or equal to $2r$.

Once a channel containing the solution path is found, the shortest path in the channel can be efficiently computed with the Funnel algorithm by extending it to handle clearance, as illustrated in Figure 4.5. The algorithm performs equivalently to its original version. Let p be the starting point of the path and u and v be the polygon vertices at the extremities of the funnel (or its *door*). The notation π_r is now used to denote a path that is composed of circle arcs and tangents of distance r to the polygon edges. Paths $\pi_r^*(p, v)$ and $\pi_r^*(p, u)$ may travel together for a while, and at some vertex a they diverge and are concave until they reach the circles centered at u and v. The funnel is the region delimited by segment uv and the concave chains $\pi_r^*(a, v)$ and $\pi_r^*(a, u)$, and a is its apex. The vertices of the funnel are stored in a double-ended queue Q for efficient processing.

Figure 4.5 illustrates the insertion process of a new vertex w. Points from the v end of Q are popped until b is reached, which is the vertex that will maintain a concave chain to w. If the apex of the funnel is popped during the process, it becomes part of the path so far and the funnel advances until reaching the destination point q. When q is reached it will be then connected to either the apex or one of the boundaries of the funnel in order to finally compose the shortest path, in the same manner as in the original funnel algorithm. In this version with clearance, when clearance values are relatively large it is possible that a new internal turn collapses the boundaries of the funnel. Such situation does not affect the overall logic of the algorithm; however, a specific correction test has to be included each time a new apex a is reached.

In summary, LCTs efficiently support the computation of paths with arbitrary clearance and the representation can thus be shared by agents with arbitrary clearance requirements. LCTs do not require the computation of the medial axis and are able to represent clearance information with less nodes than the medial axis. Figure 4.6 illustrates a LCT of the same environment shown in the CDT of Figure 4.2. It is possible to notice that the long boundary edges have been refined in order to enable the local clearance property. Additional properties, benchmarks, and algorithms for dynamic updates and robust handling of self-intersecting obstacles are described in the original LCT publication of [Kallmann, 2014].

4.2   DESIGNING NAVIGATION MESHES

The geometric representations discussed in the previous section can all be used as the underlying structure of a navigation mesh. Before selecting a geometric representation, it is important to identify the types of navigation areas that have to be represented and what types of queries will be needed in a given virtual world.

The process of constructing a navigation mesh can be subdivided in two main phases: detection and extraction of the navigable surfaces in a given 3D virtual world, and cell decomposition and representation of the surfaces as an interconnected navigation mesh. In the next sections the

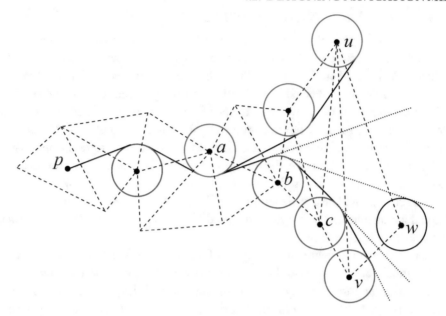

Figure 4.5: The r-funnel algorithm. The red circles are centered at the top vertices of the funnel and the blue circles are centered at the bottom vertices.

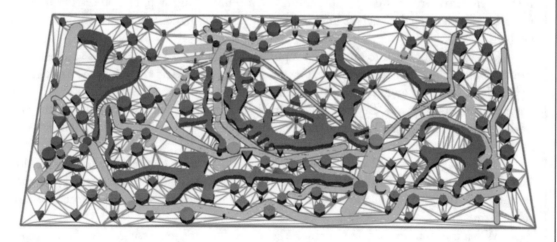

Figure 4.6: Local Clearance Triangulations have $\mathcal{O}(n)$ triangles and are computed with refinement operations on a CDT of a given environment. The achieved local clearance property allows LCTs to be shared by agents of different path clearance requirements.

overall process of building a navigation mesh and selecting suitable methods and representations is discussed.

4.2.1 EXTRACTING NAVIGATION MESHES FROM 3D WORLDS

A number of steps have to be taken into account in order to extract navigable surfaces from a given environment. First of all, the navigation capabilities of the agents to be simulated have to be clearly specified. Based on these capabilities, surfaces that are acceptable for navigation can be then detected and finally connected to each other according to the chosen representations.

The most common case involves the specification of limits related to usual human-like locomotion behaviors, such as: the maximum step height that agents can accommodate when climbing stairs or when walking over small obstacles, the maximum terrain slope that agents can accept when navigating on a surface, the minimum height that agents require for being able to pass under obstacles, the maximum jumping distance that agents can overcome with a jumping behavior (when available), etc. Clearly, depending on the scenario and application at hand, a number of additional parameters can be considered, and different types of agents may be associated with different sets of navigation limits.

When agents can navigate different types of terrains, such as water, grass, or pavement, such information should be annotated in the virtual word so that the boundaries between the different types of terrain can be automatically detected and represented. These regions will generally lead to varied traversal costs to be taken into account during path planning, and to locomotion behavior transition points annotated in the navigation mesh.

Given the collection of parameters specifying navigation capabilities, the environment can then be processed automatically. Although it is a pre-processing step, fast processing times are important for allowing designers to interactively edit the environment until achieving their design goals.

The typical approach to analyze an input 3D virtual world is to rely on a volumetric decomposition of the whole space occupied by the scene. A volumetric analysis is the most generic approach for handling an environment described without any guarantees on the connectivity or correctness of its polygons. Because it is desirable to not impose any restrictions on the work of designers, the processing has to be robust with respect to degeneracies in the models such as interpenetrating geometry, gaps, etc.

The popular Recast software [Mononen, 2015] illustrated in Figure 4.7 relies on a voxelization of the scene, which is then partitioned with a cell and portal identification method [Haumont et al., 2003]. It is also possible to use GPU techniques to quickly voxelize and process an input scene [Oliva and Pelechano, 2013].

When the input scene does not require generic volumetric processing, specialized methods operating directly on the input geometry can also be developed. For example, Lamarche [2009] projects polygons from different layers in the lowest layer to then compute a subdivision that encodes the heights of the layers above it. The information allows to determine navigable regions with respect to height constraints.

Once the navigable surfaces are identified, they are then converted into a unified navigation mesh representation. At this point semantic information relative to special navigation or access features can be considered. For example, doors and elevators will create connections (graph edges) between surfaces, and the connections can be later turned on or off at run-time.

An example of a typical special navigation capability that may connect disconnected layers are jumps. A jump can be specified as a simple behavior able to overcome small obstacles, or as a complex behavior that can connect relatively distant layers in varied relative positions. Given a jump specification, layers that can be connected by the jump are typically augmented with special links specifying the connection, these are often called *off-mesh links*.

An important step during the computation of an unified navigation mesh is to choose the polygonal cell decomposition scheme used on each layer. The result is often a multi-layer representation with each layer represented in a chosen polygonal cell decomposition scheme. Most of the approaches are developed as planar decompositions which are extended to connect different layers [Oliva and Pelechano, 2013, van Toll et al., 2011]. Layers can be processed in the plane and then be projected to non-planar surfaces when needed.

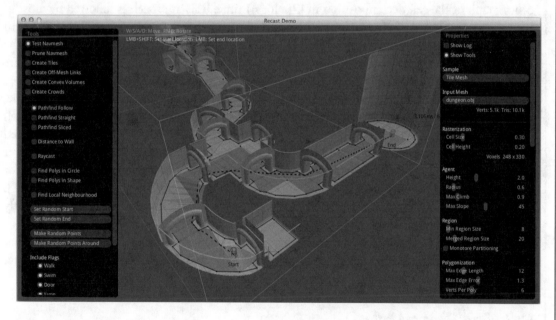

Figure 4.7: Main screen of the Recast navigation mesh construction software [Mononen, 2015]. Several parameters can be specified in order to identify navigable surfaces in an environment and to compute a global navigation mesh that can be used for path planning. The underlying representation is a triangulation with fixed clearance from obstacles. Copyright Mikko Mononen. Reproduced with permission.

4.2.2   SELECTING A CELL DECOMPOSITION SCHEME

The chosen polygonal cell decomposition scheme for the navigable layers of the environment will play a key role on the properties, efficiency, and types of navigation and path planning queries that can be handled. Basically, any of the cell decomposition schemes described in the previous chapters can be employed, and additional methods can also be found in the research literature. The key factors to observe when selecting a cell decomposition scheme are summarized below.

Linear Number of Cells

In order to ensure maximum efficiency, a navigation mesh layer should represent the environment with $\mathcal{O}(n)$ number of cells in order to allow search algorithms to operate on the structure at optimal running times. As in earlier chapters, n denotes the total number of vertices used to describe the planar polygonal obstacles in the given layer.

A linear number of cells will allow graph search algorithms to typically run on the adjacency graph of the cell decomposition in $\mathcal{O}(n \log n)$ time. This approach is followed by most of the navigation meshes used in practice. Although the search time can be reduced to $\mathcal{O}(n)$ with specialized planar graph search algorithms [Klein et al., 1994], the extra overhead including the needed pre-computation are factors that negatively influence practical performances, and that are difficult to handle in situations where the environment may dynamically change.

Optimizations are also possible to reduce the number of cells to a minimum. For example, it is possible to build a higher-level adjacency graph connecting only the degree-3 cells, which are the junction cells that connect 3 or more corridors. Such a higher level graph can be encoded in the structure with additional links allowing search algorithms to only visit a reduced number of cells. Another optimization is to reduce the number of cells by relying on larger cells. For example, the Neogen approach is based on large almost-convex cells [Oliva and Pelechano, 2011]. The drawback is that there is less resolution to encode information in the mesh or to efficiently retrieve properties, such as arbitrary clearance, from the mesh.

As seen in the previous chapters, choosing to rely on a linear size cell decomposition that is independent of the query points means that there are no straightforward methods to extract globally shortest paths from the representation.

Optimality of Computed Paths

Computing globally shortest paths in the plane, or Euclidean shortest paths (ESPs), directly from a generic cell decomposition is not a straightforward task. Either specialized search methods that take at least quadratic running time have to be employed [Kapoor et al., 1997], or specialized structures, such as the visibility graph, the SPT, or the SPM (Chapter 3), have to be computed and maintained.

The typical approach for when ESPs are necessary is to build and search the *visibility graph* [de Berg et al., 2008, Lozano-Pérez and Wesley, 1979, Nilsson, 1969] of the environment, however taking at least quadratic time to both build and search the graph. Visibility graphs are also difficult to be efficiently maintained in dynamic scenarios because of the possible high number of edges, and also because a single vertex may be connected to all the others. Despite these difficulties, visibility graphs still represent the most direct approach for computing shortest paths in the plane.

Although alternatives exist for computing ESPs, most navigation applications in virtual worlds do not impose the computation of globally shortest paths as a requirement. Fast, simple, and robust approaches are often preferred, and the $\mathcal{O}(n \log n)$ path computation time with standard graph search algorithms has been the approach of choice.

A navigation mesh should however facilitate the computation of quality paths. If ESPs cannot always be found, other guarantees on the type of paths that are computed should be provided. A reasonable expectation is that locally shortest paths should be efficiently computed, and additional characterizations related to quality may be adopted. Cell decomposition approaches that are triangulations or that can be easily reduced to triangulations offer a fast approach for computing locally shortest paths with the use of the Funnel algorithm (Figures 3.2 and 4.5).

Paths with Arbitrary Clearance

Clearance is an important aspect of navigation and a navigation mesh should provide an efficient mechanism for computing paths with arbitrary clearance from obstacles. This means that the structure should not need to know in advance the clearance values that will be used. A weaker and less desirable way to address clearance is to pre-compute information specifically for each clearance value in advance.

The most complete approach for addressing clearance is to explicitly represent the medial axis of the environment [Bhattacharya and Gavrilova, 2008, Geraerts, 2010]. LCTs (Section 4.1.4) do not encode the medial axis and instead offer a triangular mesh decomposition that carries just enough clearance information to be able to compute paths of arbitrary clearance, without the need to represent the intricate shapes the medial axis can have. If a path of maximum clearance is required, the medial axis of a triangulated path corridor can still be computed in linear time with available algorithms [Chin et al., 1999]. Alternatively, additional clearance from obstacles can be added to a path by pushing the path away from nearby obstacles as much as desired while there is clearance. Efficient implementations of such a procedure will typically perform conservative tests and thus will not exactly converge to the medial axis. However, situations where the exact medial axis would be needed are not commonly found in practice.

Specific pre-computation per clearance value is usually needed when clearance is addressed by cell decompositions not specifically designed to capture clearance in all narrow passages of the environment. For example, in the Neogen approach the larger cells require specific computations at the portals for each clearance value to be considered [Oliva and Pelechano, 2013].

Representation Robustness

A navigation mesh should be robust to degeneracies in the description of the environment. This aspect is first handled during the volumetric extraction of the navigable surfaces in the virtual world, but robustness issues may still arise at the planar level both during construction time and during run-time operations.

It is well known that the limited precision of floating point operations are often not sufficient for achieving robustness in geometric computations. One approach is to rely on arbitrary precision representation for all geometric computations, however imposing a significant performance penalty on the final system. Certain specific operations can be implemented robustly with the use of exact geometric predicates [Devillers and Pion, 2003, Shewchuk, 1997].

Robustness becomes particularly difficult when obstacles are allowed to be removed and inserted in the navigation mesh at run-time. When obstacles are inserted, undesired self-intersections and overlaps may occur, and intersection points computed with floating point operations may not exactly lie on the intersecting lines. Such imprecision eventually leads to vertices placed at illegal locations. Being robust is crucial for allowing dynamic updates to occur, in particular when users are allowed to make arbitrary updates at run-time.

An approach for handling robust dynamic updates that can be extended to any type of triangulation has been proposed as part of the LCT approach [Kallmann, 2014]. The solution is based on fast floating point arithmetic and relies on a carefully designed combination of robustness tests, one exact geometric predicate, and adjustment of illegal vertex coordinates.

Dynamic Updates

A navigation mesh should be able to efficiently update itself to accommodate dynamic changes in the environment. Dynamic updates are crucial for supporting many common events that happen in virtual worlds. Updates can reflect large changes in the environment or small ones, such as doors opening and closing, or agents that stop and become obstacles for other agents.

In general, all approaches for navigation meshes can be extended to accommodate dynamic operations. The general trade-off is that the more complex the structure is, the more complex and expensive it is to maintain it dynamically. For instance, there are several hierarchical representations that are possible for speeding up path search; however, if a navigation mesh is associated with a hierarchical structure the hierarchy has also to be updated for every change in the navigation mesh.

The overall chosen approach to address dynamic updates should first take into account how often path queries and dynamic updates are executed, and the correct representations and methods should then be determined accordingly. For instance, if dynamic updates occur often but path queries not so often, dynamic updates can be delayed until the next path query, and depending on the number and extent of the required updates, a re-build of the affected area might be preferable. Such considerations are important and become considerably dependent on the application.

PART III

Advanced Planning Techniques

CHAPTER 5

Extending Basic Search Techniques

Parts I and II of this book reviewed the several foundational techniques for path search and for designing discrete representations for efficient path planning. While the foundational techniques are always present, there are a number of additional challenges to be addressed when designing planning solutions for interactive virtual worlds. The remaining chapters of this book review several techniques that can be used for addressing these additional challenges. Of particular importance are planning techniques that are able to generate solutions with strict time constraints in large and complex problem domains. While the presentation is focused on describing the work of the authors on these areas, broad discussions on the state of the art are also included.

There are three main areas that planning algorithms have to address in order to address the needs of modern interactive virtual worlds. (1) Planners must satisfy strict time constraints and must be able to efficiently repair solutions for dynamically changing environments. (2) Planners should be able to enforce spatial as well as temporal constraints that influence the type of paths that are produced. (3) To keep pace with the ever-growing size of game worlds, planning algorithms must be able to scale to large environments and to many simulated agents.

In Part III we review several advanced planning techniques for real-time planning in dynamic problem domains. In the next sections we review planning techniques that leverage the use of parallelization, abstractions, and hierarchies to support real-time applications. To address challenge (1), Chapter 5 describes extensions to the classical A* algorithm and describes *anytime* and *dynamic* extensions which enforce strict time constraints, while efficiently handling dynamic environments. To address challenge (2), Chapter 6 proposes extensions to existing anytime dynamic planners in order to enforce constraints on the resulting plans that are produced. To address challenge (3), Chapter 7 presents new approaches for leveraging massive parallelization while enforcing the same properties as serial planners, in order to enable real-time multi-agent planning in large, complex, and dynamic environments.

5.1 BACKGROUND

This section describes the use of parallelization, hierarchical abstractions, and local methods for real-time planning.

5.1.1   PARALLEL SEARCH

There are existing efficient search techniques based on expanding trees that are able to exploit multiple processors [Ferguson and Korf, 1988, Powley et al., 1993], Tree-based search algorithms such as breadth-first search techniques [Powley et al., 1993] and alpha-beta pruning [Ferguson and Korf, 1988] have been optimized to exploit multiple processors. However, these methods are not well suited for massive parallelization on graphics hardware because these approaches cannot be reduced to simple arithmetic operations which can be performed efficiently on GPU threads.

Wavefront-based algorithms [Pal et al., 2011] are particularly suitable for massive parallelization as each environment subdivision can independently monitor the propagation of the wave which can be reduced to a simple operation that requires only local memory accesses and can be efficiently performed on GPU threads, one for each environment subdivision (e.g., a grid cell). These approaches have been demonstrated in a wide variety of problem domains [Hoisie et al., 1998], to yield substantial performance benefits over serial algorithms. GPU accelerated path planning algorithms [Fischer et al., 2009] provide tremendous performance boost, enabling the solution of higher dimensional problems, but return sub-optimal paths. The work in [Zhou and Zeng, 2015] proposes a massively parallel variant of the classical A* search algorithm. The work in [Delling et al., 2011] demonstrates shortest path calculations for graph based searches on the GPU. [Buluc et al., 2010] uses a blocked recursive elimination strategy to utilize the computational power and memory bandwidth of the GPU. Randomized searches [Gini, 1996, Kider et al., 2010] have been successfully ported to the GPU, by doing multiple short-range searches in parallel, but provide no optimality guarantees. Distance fields can be used to solve multi-agent planning on the GPU [Torchelsen et al., 2010]. Crowd simulation techniques [Bleiweiss, 2009, Guy et al., 2009] exploit GPU hardware to accelerate local collision avoidance for crowds but does not handle global planning.

5.1.2   HIERARCHICAL SEARCH

One possible way for reducing planning time is by creating different levels of abstraction over the state space, and planning in the lower-dimensional abstract search space to reduce computational complexity while sacrificing optimality guarantees. The idea of clustering states together to create a hierarchical representation of the environment is a very effective way of reducing computation costs at the expense of solution quality. The work in [Sturtevant, 2007] proposes a memory efficient representation based on splitting the state space into sectors and regions. At the expense of sacrificing optimality, the method greatly outperforms classical A*. A hierarchical approach to task and motion planning which does not require discretization of the state space is shown in [Kaelbling and Lozano-Pérez, 2011]. HAA* [Harabor and Botea, 2008] is a hierarchical strategy which creates abstractions that can be used by heterogeneous agents with different capabilities.

HPA* [Botea et al., 2004] was introduced as a planner that is able to perform efficient on-line search. It is a hierarchical method that finds solutions to several smaller sub-problems, called intra-edge paths, and utilizes these to construct a solution to the original problem. The approach has a high initial cost to construct abstract states (clusters of states). Once this pre-processing step is completed, online planning is very efficient. However, one drawback of this method is that it is limited to static domains. This approach is extended to dynamic domains in [Kring et al., 2010] to DHPA*. This method keeps a cache of every intra-edge path for every abstract state and after changes in the environment are observed, the solutions to each affected cluster are recomputed. Even though DHPA* limits the re-computation to the affected clusters, it is still throwing away useful information within that cluster.

A different approach to create state abstractions is using quad-trees and octrees that allow for efficient traversal. The work in [Goodrich and Sun, 2005] demonstrates a method to quickly perform point location, search, and insertion operations in randomized and deterministic quad-trees by using compressed quad-trees. The work in [Frey and Marechal, 1998] describes a method of rapidly creating an adaptive mesh based on quad-trees, but is not well suited for a GPU implementation. The work presented by Li and Loew [1987a,b], provides an efficient way of performing adjacency detection based on a series of arithmetic operations performed on *quad-codes*.

5.1.3   NAVIGATION WITH CONSTRAINTS

The work of Xu and Badler [2000] describes a list of representative prepositions for constraining motion trajectories in goal-directed navigation. The work of André et al. [1986] analyses the semantics of spatial relations, including "along" and "past," to characterize the path of moving objects. In addition, several search methods based on homotopy classes have been proposed. Bhattacharya et al. [2012a,b] explore the use of homotopy classes of trajectories in graph-based search for path planning with constraints. This method is extended by Bhattacharya et al. [2012a] to handle 3D spaces. A homotopy class-based approach to A*, called *HA** [Hernandez et al., 2011], ensures optimality and directs the search by exploring areas that satisfy a given homotopy class. The work of Phillips et al. [2013] demonstrates constrained manipulation using experience graphs. The works in [Geraerts, 2010, Kallmann, 2014] embed additional information in the underlying environment representation in order to efficiently compute locally shortest paths with clearance constraints. Sturtevant [2009] investigates the feasibility of introducing motion constraints in pathfinding and proposes the use of perimeter heuristics and intermediate search truncation to provide and order of magnitude speedup in certain problem domains. Follow-up work by [Sturtevant, 2013] explores the integration of human relationships into path planning in games, incorporating "relationship distance" and line of sight as positive and negative weights in the path planning problem.

5.1.4   POTENTIAL FIELDS

The approach of potential fields [Arkin, 1987, Goldenstein et al., 2001, Shimoda et al., 2005, Warren, 1989, 1990] generates a global field for the entire landscape where the potential gradient is contingent upon the presence of obstacles and distance to the goal. These methods suffer from local minima where the agents can get stuck and never reach the goal. Since a change in target or environment requires significant re-computation, these navigation methods are generally confined to systems with non-changing goals and static environments. Dynamic potential fields [Treuille et al., 2006] have been used to integrate global navigation with moving obstacles and people, efficiently solving the motion of large crowds without the need for explicit collision avoidance. The work of Kapadia et al. [2009b, 2012] uses local variable-resolution fields to mitigate the need for computing uniform global fields for the whole environment, and uses best-first search techniques to avoid local minima.

5.1.5   CONTINUOUS WEIGHTED-REGION NAVIGATION

Various continuous navigation methods in non-uniformly-weighted space have been developed. One continuous Dijkstra technique finds shortest paths using the principles of Snell's law [Mitchell and Papadimitriou, 1991]. Methods have been shown for finding exact solutions to simpler weighted-region problems [Narayanappa et al., 2005]. Suboptimal solutions can be found using an efficient approximation algorithm [Reif and Sun, 2000].

5.1.6   LOCAL COLLISION AVOIDANCE

There is extensive work [Pelechano et al., 2008] that relies on local goal-directed collision-avoidance for simulating large crowds. These approaches rely on locally optimal solutions for simulating the movement of hundreds, even thousands of autonomous agents at interactive rates, and rely on a global static path as a reference trajectory to move along, while avoiding dynamic threats such as neighboring agents. For comprehensive surveys on local collision avoidance we refer the readers to other specific texts on the topic such as [Kapadia and Badler, 2013, Kapadia et al., 2015b, Thalmann and Musse, 2007].

5.1.7   REAL-TIME PLANNING IN DYNAMIC PROBLEM DOMAINS

There are two main requirements that planners must satisfy before they can be used in real-time practical applications. First, these approaches must be able to satisfy strict time constraints which are true for real-time applications. Second, these approaches must efficiently operate in dynamic problem domains where the search graph may be constantly changing. To meet these requirements, we present a broad overview of recent developments in planning that are suitable for real-time dynamic applications. Section 5.2 reviews anytime approaches to planning that meet strict time constraints, and dynamic planners that are designed to efficiently reuse previous searches in order to quickly repair solutions to accommodate dynamic problem domains. Section 5.3 presents Anytime Dynamic Search, or AD* [Likhachev et al., 2005] which combines the properties of anytime and dynamic search methods.

5.2 EXTENSIONS TO CLASSICAL A*

Discrete search algorithms such as A*, described in detail in Chapter 2, find the least cost path from the start to the goal while using a heuristic estimate of the distance to the goal to focus and guide the search. Dijkstra [1959] does not rely on the heuristic and uniformly expands states in all directions until eventually reaching the goal state. A* relies on the heuristic to focus the exploration of states. However, for large problems A* still needs to expand many nodes and can very quickly run out of memory. To offset these limitations, several variants of the classical A* search algorithm have been proposed in the community which are described below.

5.2.1 WEIGHTED A* SEARCH

A simple extension [Ira Pohl, 1973] is introduced which weighs the influence of the heuristic on the search using an inflation factor, ϵ. For $\epsilon > 1$, it increases the bias of the heuristic on state expansion where states that are more promising candidates to the goal are given preference for expansion. Weighted A* trades off optimality for speed and produces ϵ−suboptimal solution such that the cost of the resulting path produced $g_w < \epsilon g_{opt}$ where g_{opt} is the cost of the optimal path generated using A*. Weighted A* has been shown to produce orders of magnitude speedup in many, more complex planning domains. However, it does require the careful modeling of heuristic function which can greatly influence search performance.

5.2.2 ANYTIME SEARCH

Anytime Repairing A*, or ARA* [Likhachev et al., 2003] is an extension with anytime properties, which can quickly generate a sub-optimal solution while satisfying time constraints, and gradually converge to an optimal while reusing previous search efforts. ARA* performs a series of repeated weighted A* searches while iteratively decreasing ϵ to first quickly find a highly suboptimal solution. And then if time permits, it iteratively improves the solution by reducing ϵ and reusing previous plan efforts to accelerate subsequent searches.

The key to reusing previous plan efforts is keeping track of inconsistent states, or more specifically over-consistent states. A state s is locally inconsistent if its g-value, $g(s)$ decreases, where $g(s)$ is the actual cost of reaching the state, s from the start state. Suppose the state s is the best predecessor for some state s'. If $g(s)$ decreases then the current estimated cost of $g(s')$ is greater than its actual cost and is thus overconsistent. The path calculations for ARA* are modified to keep track of all inconsistent states. At every plan stage it initializes the open list, which is the list of non-visited nodes, in order to include all the inconsistent states to make sure they are expanded in the current plan iteration. Whenever an inconsistent state is expanded, it corrects its inconsistency and in turn makes its successors inconsistent. In this way, the local inconsistency is propagated till the resulting solution returned is guaranteed to be consistent. For more details on ARA* we refer the reader to [Likhachev et al., 2003].

5.2.3 DYNAMIC SEARCH

A large variety of algorithms [Barbehenn and Hutchinson, 1995, Ersson and Hu, 2001, Koenig et al., 2004, Ramalingam and Reps, 1996] have been proposed in the robotics community for efficiently repairing existing plans (re-planning) to accommodate dynamic search graphs. Dynamic A* (D*) [Stentz, 1995] and D* Lite [Koenig and Likhachev, 2002] are popular approaches for dynamic planning, and have been shown to be up to two orders of magnitude faster than classical A* for dynamic problems. We refer the readers to [Koenig and Likhachev, 2002] and provide a brief review of D* Lite below.

D* Lite maintains a least-cost path from a start state to a goal state, with $g(s)$ estimating the cost from each state s to the goal. Note that the definition of $g(s)$ is different here, since D* Lite performs a backward search from the goal state to the start state. It also stores a one-step lookahead value $rhs(s)$, which potentially provides a more informed estimate of $g(s)$ by choosing the predecssor, s' which has the minimum cost. It is thus computed as $rhs(s) = g(s') + c(s', s)$, where $c(s, s')$ is the cost of the transition from s to s'. A state is called consistent iff $g(s) = rhs(s)$. A state is overconsistent if $g(s) > rhs(s)$ or underconsistent if $g(s) < rhs(s)$. Similar to A*, D* Lite uses a heuristic and a priority queue to focus its search and to order its cost updates efficiently. The priority queue OPEN always holds exactly the inconsistent states; these are the states that need to be updated and made consistent. A lexicographic ordering **key**(s) is used to order nodes in OPEN which is described in more detail in Section 5.3. D* Lite expands states from the queue in increasing priority, updating $g(s)$ and $rhs(s)$, until there is no state in the queue with a key value less than that of the start state. Its execution is thus identical to a backward A* search during the first plan execution. If the search graph changes, it updates the rhs-values of each state immediately affected by the change and adds inconsistent states to OPEN. It then repeats the same process to efficiently repair the path by resolving all inconsistent nodes in the queue.

5.3 ANYTIME DYNAMIC SEARCH

Anytime D* combines the properties of dynamic search and ARA* to provide a planning solution that meets strict time constraints and is able to efficiently update its solutions to accommodate dynamic changes in the environment. We provide a brief overview of the algorithmic details below and refer the readers to [Likhachev et al., 2005] for more details.

5.3.1 PLAN COMPUTATION

Similar to ARA*, it performs a series of repeated searches by iteratively decreasing the inflation factor. This allows it to quickly generate a suboptimal solution while meeting strict time constraints and iteratively refine the solution if time permits. It interleaves planning with execution by moving the agent along the current published path. It can also efficiently repair plans to accommodate world changes as well as start movement which allows it to interleave planning with execution. It does this by monitoring world changes which may make current expanded states inconsistent which are then resolved in subsequent plan iterations. If significant world changes are observed, it can choose to increase the inflation factor to expedite the search to quickly generate a sub-optimal solution. In some cases it may be more efficient to plan from scratch.

Procedure **ComputeOrImprovePath** (Algorithm 10) is invoked each time the planning task is executed, and caters to underconsistent states that may arise in the event of dynamic changes. This function monitors events and calls the appropriate event handlers for changes in start, goal, and constraints. Given a maximum amount to deliberate t_{max}, it refines the plan and publishes the ϵ-suboptimal solution using the AD* planning algorithm [Likhachev et al., 2005]. The method **UpdateState**(s) is defined in Algorithm 11.

Algorithm 10 - Compute or Improve Path

Input: The maximum deliberation time, t_{max}

1: **ComputeOrImprovePath** (t_{max})
2:   **while** ($\min_{s \in \text{OPEN}}(key(s)) < key(s_{goal}) \vee rhs(s_{goal}) \neq g(s_{goal}) \vee \Pi(s_{start}, s_{goal}) =$ NULL) $\wedge\, t < t_{max}$ **do**
3:     $s = \arg_{s \in \text{OPEN}} \min(key(s))$;
4:     **if** $(g(s) > rhs(s))$ **then**
5:       $g(s) = rhs(s)$;
6:       CLOSED = CLOSED $\cup\, s$;
7:     **else**
8:       $g(s) = \infty$;
9:       **UpdateState**(s);
10:    **end if**
11: **end while**

AD* performs a backward search and maintains a least cost path from the goal s_{goal} to the start s_{start} by storing the cost estimate $g(s)$ from s to s_{goal}, similar to D* Lite. However, in dynamic environments, edge costs in the search graph may constantly change and expanded nodes may become inconsistent. Hence, a one-step look ahead cost estimate $rhs(s)$ is introduced [Koenig and Likhachev, 2002] to determine node consistency.

The priority queue OPEN contains the states that need to be expanded for every plan iteration, with the priority defined using a lexicographic ordering of a two-tuple **key**(s), defined for each state, which is defined in Algorithm 12. OPEN contains only the inconsistent states ($g(s) \neq rhs(s)$) which need to be updated to become consistent.

An increase in edge cost may cause states to become under-consistent ($g(s) < rhs(s)$) where states need to be inserted into OPEN with a key value reflecting the minimum of their old cost and their new cost. In order to guarantee that under-consistent states propagate their new costs to their affected neighbors, their key values must use uninflated heuristic values. This means that different key values must be computed for under- and over-consistent states, as shown

---

Algorithm 11 - State Update

Input: The state s

1: **UpdateState**(s)
2: **if** $(s \neq s_{\text{start}})$ **then**
3:     $s_n = \arg_{s_n \in \text{pred}(s)} \min(c(s, s_n) \cdot M_{\mathbf{C}}(s, s_n) + g(s_n));$
4:     $rhs(s) = c(s, s_n) \cdot M_{\mathbf{C}}(s, s_n) + g(s_n);$
5:     $prev(s) = s_n;$
6: **end if**
7: **if** $(s \in \text{OPEN})$ remove s from OPEN;
8: **if** $(g(s) \neq rhs(s))$ **then**
9:     **if** $(s \notin \text{CLOSED})$ insert s in OPEN with key(s);
10:     **else** insert s in INCONS;
11: **end if**
12: Insert s in VISITED;

Algorithm 12 - Key definition

Input: The state s
Output: The 2-tuple key value for state s

1: **key**(s)
2: **if** $(g(s) > rhs(s))$ **then**
3:     **return** $[rhs(s) + \epsilon \cdot h(s, s_{\text{goal}}); rhs(s)];$
4: **else**
5:     **return** $[g(s) + h(s, s_{\text{goal}}); g(s)];$
6: **end if**

---

in Algorithm 12. This key definition allows AD* to efficiently handle changes in edge costs and changes to inflation factor.

Nodes are expanded in increasing priority until there is no state with a key value less than the start state. A heuristic function $h(s, s_n)$ computes an estimate of the optimal cost between two states, and is used to focus the search toward s_{start}.

Instead of processing all inconsistent nodes, only those nodes whose costs may be inconsistent beyond a certain bound, defined by the inflation factor ϵ are expanded. It performs an initial search with an inflation factor ϵ_0 and is guaranteed to expand each state only once. An INCONS list keeps track of already expanded nodes that become inconsistent due to cost changes in neighboring nodes. Assuming no world changes, ϵ is decreased iteratively and plan quality is improved until an optimal solution is reached ($\epsilon = 1$). Each time ϵ is decreased, all states made inconsistent due to change in ϵ are moved from INCONS to OPEN with **key**(s) based on the reduced inflation factor, and CLOSED is made empty. This improves efficiency since it only expands a state at most once in a given search and reconsidering the states from the previous search that were inconsistent allows much of the previous search effort to be reused, requiring only a minor amount of computation to refine the solution.

When change in edge costs are detected, new inconsistent nodes are placed into OPEN and node expansion is repeated until a least cost solution is achieved within the current ϵ bounds. When the environment changes substantially, it may not be feasible to repair the current solution and it is better to increase ϵ so that a less optimal solution is reached more quickly.

5.3.2 DYNAMIC EVENTS

AD* uses a backward search to handle agent movement along the plan by recalculating key values to automatically focus the search repair near the updated agent state. It can handle changes in edge costs due to obstacle and start movement, and needs to plan from scratch each time the goal changes. Algorithm 13 monitors events in the simulation and triggers appropriate routines, which are described below.

StartChangeUpdate

When the start moves along the current plan, the key values of all states in OPEN are recomputed to re-prioritize the nodes to be expanded. This focuses processing toward the updated agent state allowing the agent to improve and update its solution path while it is being traversed. When the new start state deviates substantially from the path, it is better to plan from scratch. Algorithm 14 provides the routine to handle start movement.

GoalChangeUpdate

Algorithm 15 clears plan data and resets ϵ whenever the goal changes and plans from scratch at the next step.

ConstraintChangeUpdate

Chapter 6 presents an approach of how AD* can be employed to efficiently handle dynamic constraints that can be used to influence and control path calculations.

Algorithm 13 - Event Handler

Input: Event trigger signal, SIGNAL

  1: **EventHandler**(SIGNAL)
  2: **if** (START_CHANGED) **then**
  3:    **StartChangeUpdate** (s_c);
  4: **end if**
  5: **if** (GOAL_CHANGED) **then**
  6:    **GoalChangeUpdate** (s_{new});
  7: **end if**
  8: **if** (CONSTRAINT_CHANGED) **then**
  9:    **for each** constraint change c **do**
10:      **ConstraintChangeUpdate** (c, $\vec{x}_{\text{prev}}, \vec{x}_{\text{next}}$);
11:    **end for**
12: **end if**

Algorithm 14 - Updates due to change in start state

Input: New start state, s_c

  StartChangeUpdate(s_c)
  if ($s_c \notin \Pi(s_{\text{start}}, s_{\text{goal}})$) **then**
    ClearPlanData();
    $\epsilon = \epsilon_0$;
  else
    $s_{\text{start}} = s_c$;
    for each $s \in$ OPEN **do**
      Update **key**(s);
    end for
  end if

Algorithm 15 - Updates due to change in goal state

Input: New start state, s_{new}

  GoalChangeUpdate(s_{new})
  ClearPlanData()
  $\epsilon = \epsilon_0$;
  $s_{\text{goal}} = s_{new}$;

CHAPTER 6

Constraint-Aware Navigation

This chapter extends the anytime dynamic planner described in Section 5.1.7 and introduces a planning approach for constraint-aware navigation that enables autonomous agents to be more aware of the semantics of objects in the environment and thus interpret high-level navigation goals with dynamic and meaningful spatial path constraints. At a personal scale, such constraints may include: specific location investigation ("check behind the building"), implemented as a goal state; dynamic agent evasion and stealth ("avoid being seen") and organization ("stay between these two guys"), both implemented using dynamic constraints; or instructions with path requirements for the mission ("follow the road"), implemented using obstacles or a series of goals. Our approach scales up to larger scenarios as well, allowing constraints such as neighborhood awareness ("avoid this part of the map") or time-of-day awareness ("stay on main roads at night"), implemented using time-dependent constraint weights.

Our initial hybrid discretization of the environment combines the computational benefits of triangulations with a dense uniform grid, ensuring sufficient resolution to account for dynamic constraints. Planning results using this hybrid discretization are highly dependent upon the quality of the environment triangulations. In order to create a suitable navigation graph, we devise an "Adaptive Highway" constraint-dependent planning domain, which has an increased branching factor (and therefore denser environment discretization) in areas affected by constraints.

Hard constraints, which must be satisfied, effectively prune invalid transitions in the search graph (using infinite transition costs), while soft constraints (attractors or repellers) have a multiplicative effect on the cost of choosing a transition. Constraints are represented as continuous potential-like fields, which can be easily superimposed to calculate the cumulative effect of multiple constraints in the same region, and can be efficiently queried during search exploration. This chapter summarizes our prior work, [Kapadia et al., 2013d, Ninomiya et al., 2014], on this topic.

We demonstrate the potential for the approach by presenting challenging navigation problems in complex environments using combinations of constraints on static obstacles and dynamic agents, including various combinations of hard, soft, attracting, and repelling constraints.

6.1 PROBLEM DEFINITION

The problem domain $\Sigma = \langle \mathbf{S}, \mathbf{A} \rangle$ defines the set of all possible states $\mathbf{S}$, and the set of permissible transitions $\mathbf{A}$. Every problem instance $\mathbf{P}$, for a particular domain Σ, is defined as $\mathbf{P} = \langle \Sigma, s_{\text{start}}, s_{\text{goal}}, \mathbf{C} \rangle$, where $(s_{\text{start}}, s_{\text{goal}})$ are the start and goal state, and $\mathbf{C}$ is the set of active (hard and soft) constraints. A hard constraint is used to prune transitions in $\mathbf{A}$. For example, consider a flower bed which *must* not be stepped upon. A hard constraint could be specified for that area, pruning every transition that could violate this restriction. A soft constraint influences the costs of actions in the action space, and can tend the agent toward or away from a certain region in space. A planner generates a plan, $\Pi(s_{\text{start}}, s_{\text{goal}})$, which is a sequence of states from s_{start} to s_{goal} that adheres to **C**.

6.2   ENVIRONMENT REPRESENTATION

In this section, we describe the discretized environment representation that we use for constraint-aware pathfinding, building on top of the geometric approaches that are described in Part II of this book. The challenge we encounter is as follows: a coarse resolution representation, such as a mesh-triangulation-based approach, is often used to facilitate efficient search. However, it cannot accommodate all constraints, as it has insufficient resolution in regions of the environment where constraints may be specified. Since dynamic constraints are not known ahead of time, it is impossible to simply increase triangulation density near constraints. A sufficiently dense uniformly distributed graph representation of the environment can account for all constraints (including dynamic objects), but is not efficient for large environments.

   To avoid these limitations, we explore two alternative methods of generating environment representations. First, we experimented with a "Hybrid" search graph that has sufficient resolution while still accelerating search computations by exploiting longer, coarser transitions to improve suboptimal planning speed in certain situations. More efficient results were seen with an "Adaptive Highway" domain, operating in the same *state* space as the dense environment representation but with dynamically reduced branching at states unaffected by constraint effects. This provides bounded suboptimality: the reduced-branching graph is, at worse, no more coarse than a lower-resolution dense uniform graph.

6.2.1   TRIANGULATION

We define a simple triangulated representation of the free space in the environment, represented by $\Sigma_{\text{tri}} = \langle \mathbf{S}_{\text{tri}}, \mathbf{A}_{\text{tri}} \rangle$ where elements of $\mathbf{S}_{\text{tri}}$ are the midpoints of the edges in the navigation mesh and elements of $\mathbf{A}_{\text{tri}}$ are the six directed transitions per triangle (two bi-directional edges for each pair of states). This simple triangulation can be easily replaced by other more advanced navigation meshes [Kallmann, 2014, Mononen, 2015, Oliva and Pelechano, 2011, 2013, Pettré et al., 2005, van Toll et al., 2012], and produces a low-density representation of the state and action space. Figure 6.1a illustrates Σ_{tri} for a simple environment. The triangulation domain Σ_{tri} provides a coarse-resolution discretization of free space in the environment, facilitating efficient pathfinding. However, the resulting graph is too sparse to represent paths adhering to constraints such as spatial relations to an object.

   To offset this limitation, we can annotate objects in the environment with additional geometry to describe relative spatial relationships (e.g., `Near`, `Left`, `Between`, etc.) We use these annotations to generate additional triangles in the mesh, which expands Σ_{tri} to include states

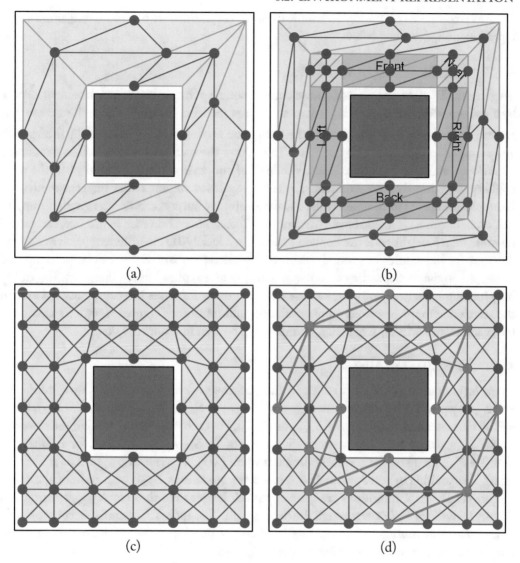

Figure 6.1: (a) Environment triangulation Σ_{tri}. (b) Object annotations, with additional nodes added to Σ_{tri}, to accommodate static spatial constraints. (c) Dense Uniform graph Σ_{dense}, for same environment. (d) A hybrid graph Σ_{hybrid} of (a) Σ_{tri} and (c) Σ_{dense}; highways (newly inserted transitions) are indicated in red.

and transitions that can represent these spatial relations. Annotations, and the corresponding triangulation, are illustrated in Figure 6.1b. These annotations are useful for constraints relative to static objects; however, Σ_{tri} cannot account for dynamic objects as the triangulation cannot be efficiently recomputed on the fly. To handle dynamic constraints, we utilize a dense, uniform graph representation, described below.

6.2.2   DENSE UNIFORM GRAPH

To generate $\Sigma_{\text{dense}} = \langle \mathbf{S}_{\text{dense}}, \mathbf{A}_{\text{dense}} \rangle$, we densely sample points in the 3D environment, separated by a uniform distance d_{grid}, which represents the graph discretization. For each of these points, we add a state to $\mathbf{S}_{\text{dense}}$ if it is close to the navigation mesh (within $\frac{\sqrt{3}}{2} d_{\text{grid}}$ of the nearest point), and clamp it to that point (Figure 6.1c). This allows the method to work well in environments with slopes, stairs, and multiple levels. While all of our environments are locally 2D, the method could be used directly for 3D domains, by testing against a volume rather than the mesh surface. Examples are shown in completely flat 2D for illustrative purposes. Since the graph is sampled in 3D, each state in $\mathbf{S}_{\text{dense}}$ could have a maximum of 26 neighbors; however, in practice, each state has no more than 8 neighbors if the domain operates in a locally 2D environment (such as navigation domains for humanoids or ground vehicles). The dense domain Σ_{dense} can be precomputed or generated on the fly, depending on environment size and application requirements. Regardless of how it is implemented, however, a dense domain greatly increases the computational burden of the search due to the increased number of nodes and transitions compared with a sparse domain (compare Figure 6.1a/c). Expansion examples can be seen in Figure 6.4a/c.

6.2.3   HYBRID GRAPH

In the first of our two attempts to mitigate the performance problem of Σ_{dense}, we combine Σ_{dense} and Σ_{tri} to generate a hybrid domain $\Sigma_{\text{hybrid}} = \langle \mathbf{S}_{\text{hybrid}} = \mathbf{S}_{\text{dense}}, \mathbf{A}_{\text{hybrid}} \approx \mathbf{A}_{\text{dense}} \cup \mathbf{A}_{\text{tri}} \rangle$. First, we add all the states and transitions in Σ_{dense} to Σ_{hybrid}. For each state in $\mathbf{S}_{\text{tri}}$, we find the closest state in $\mathbf{S}_{\text{dense}}$, creating a mapping between the state sets, $\lambda : \mathbf{S}_{\text{tri}} \rightarrow \mathbf{S}_{\text{dense}}$. Then, for each transition $(s, s_n) \in \mathbf{A}_{\text{tri}}$, we insert the corresponding transition $(\lambda(s), \lambda(s_n))$ in $\mathbf{A}_{\text{dense}}$ (if it does not already exist). The resulting hybrid domain Σ_{hybrid} has the same states as Σ_{dense} with additional transitions (Figure 6.1d). These transitions are generally much longer than those in $\mathbf{A}_{\text{dense}}$, creating a low-density network of *highways* through the dense graph. This approach is similar in concept to those of highway node routing [Schultes, 2008] and contraction hierarchies [Geisberger et al., 2008], but we exploit information provided by the navigation mesh.

In Σ_{hybrid}, a pathfinding search can automatically choose highways for long distances, and use the dense graph more heavily when the planner has additional time to compute an exact plan. As before, the dense graph allows the planner to find paths that adhere to constraints; when there is no strong influence of nearby constraints, the planner can take highways to improve its performance. In addition, with a planner like AD* [Likhachev et al., 2005], we can inflate the influence of the heuristic to very quickly produce suboptimal paths that favor highway selection, then iteratively improve the path quality by using dense transitions. This method makes it possible to maintain interactive frame rates.

6.2.4   ADAPTIVE HIGHWAY GRAPH

For practical reasons, we wish to avoid depending heavily on both the quality of the triangulations in Σ_{tri} and the behavior of the anytime dynamic planner for suboptimal performance improvements. To this end, we introduce a uniform search domain with constraint-adapting highways. This graph maintains a high density of search nodes near constraints, while automatically pruning transitions when additional accuracy is not needed. The results of this pruning approach are shown in Figure 6.3, and described here.

The basis of this domain is a lazily evaluated form of Σ_{dense}. As an example, in a 2D navigation problem, a state in Σ_{dense} has eight neighbors, and therefore eight potential transitions. However, since high graph resolution is not necessary in obstacle-free, constraint-free areas, it is possible to reduce the graph resolution in many parts of an environment. That is, the resolution of the graph should be dependent upon the absolute amount of constraint weight in that area (the *absolute-value weight field*).

In order to create this adaptive change in resolution, we emulate the "highway" transitions found in Σ_{hybrid}; however, instead of adding long-distance transitions, we can take another approach: Since $\Sigma_{adaptive}$ is aware of the expansion of the planner, it instead "skips" through nodes in a single direction. That is, from a given state (e.g., the start state), the domain will expand in eight directions, but, in the absence of constraints, each of those expanded states may have only one possible transition, which continues in the same direction. The "skipping" expansion is controlled as follows.

- When a graph state is expanded, it stores a "remaining number of skips" value into each expanded state. This keeps track of how many skips are left in a given highway, and we limit skipping expansion to (in our examples) at most four transitions in one direction.

- In addition, as the expansion continues skipping in one direction, the cumulative integral of the absolute-value weight field is also tracked. When too much absolute-value weight is encountered, we cancel skipping expansion early to ensure high graph density near constraints.

To ensure correct behavior when states are invalidated, $\Sigma_{adaptive}$ simply re-expands invalidated areas using the newly computed weight field, discarding the old expansion. The traits of the Adaptive Highway domain and performance trade-offs are described in Section 6.5.2.

Observed limitations of the current $\Sigma_{adaptive}$ might be the subject of future research, and are as follows. First, while the branching factor is correctly reduced in unweighted areas, the increased state density persists after the planner's search passes beyond a constraint. This is seen just past the positive constraint in Figure 6.3c. A potential fix would be to explicitly reduce the state density in this case. Second, the branch density is dependent upon the amount of absolute constraint weight in an area. Due to this, inside a very large, constant constraint, the graph density will still be high; better results might be achieved by adjusting branching factor based on the rate of change of the weight field. In addition, in the absence of constraints, navigation near edges of obstacles is inaccurate due to the low state density. It may be possible to remedy this by decreasing skip lengths near obstacles and walls. Finally, in Figures 6.4c/d, note that the resulting paths are very different: This is due to the fact that the Adaptive Highway domain represents the environment with less accuracy in favor of performance improvements, sometimes giving suboptimal results. However, since the skip-count in the Adaptive Highway domain is bounded, this provides a bound on the suboptimality of this method: it will never be less optimal than a domain which always takes the maximum number of skips (i.e., a domain with lower density).

6.3   CONSTRAINTS

Constraints imposed on how an agent navigates to its destination greatly influence the motion trajectories that are produced, and often result in global changes to the paths that cannot be met using existing local solutions. For example, if there is an agent who wishes to stay behind a building or outside another agent's line of sight, our method may choose very circuitous paths in order satisfy these constraints. Our framework supports hard constraints (obstacles), which must always be met; attractors, which reduce nearby transition costs; and repellers, which increase nearby transition costs.

6.3.1   PROBLEM SPECIFICATION

A constraint-aware planning problem is represented by a start state s_{start}, a goal state s_{goal}, and a set of constraints $\mathbf{C} = \{c_i\}$. Each constraint is defined as $c_i = ((\texttt{In}|\texttt{Near})\ \texttt{Annotation}_i$ with weight $w_i)$, where the weight w_i can be either positive (attracting) or negative (repelling). Hence, a specification can be written as a simple regular language:

$$\text{Move from } s_{start} \text{ to } s_{goal}\ ((\texttt{In}|\texttt{Near})\ \texttt{Annotation}_i \text{ with weight } w_i)\ *$$

Despite the simplicity of such a definition, it is important to note its flexibility: both goals and constraints can encode very rich semantics for a wide variety of planning problems. In addition, multiple problem specifications can be chained together to create more complex commands; for example, "move to the street corner, then patrol the alleyway," where "patrol" can be described as a repeating series of commands going back and forth between two points.

6.3.2   CONSTRAINT DEFINITIONS

In this section, we define the various terms and concepts used to formalize the planning problem while incorporating constraints.

Annotations
An annotation is simply a volume of space that allows the user to define the area of influence of a constraint. These volumes can be defined in real time, and may be dynamically calculated from

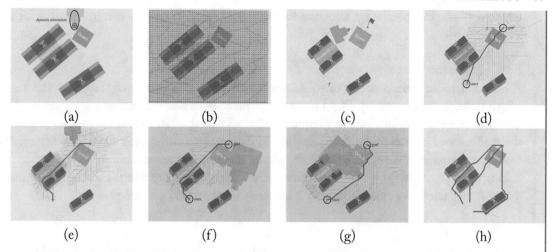

Figure 6.2: (a) An environment with annotations such as Front, Left, Between, and a simple LineOfSight. (b) Transitions in hybrid domain Σ_{hybrid}. (c) A specific problem instance with the following constraints: Not In Grass $\wedge$ Not Near LineOfSight (Agent) $\wedge$ Near Between (B, C). (d) Static optimal path, in absence of constraints. (e) Resulting path produced for problem instance (c). (f)–(g) Plan repair to accommodate moving LineOfSight constraint. The Between constraint is invalidated due to the LineOfSight constraint, of higher priority. (h) Multiple characters simultaneously navigate under different constraint specifications, producing different paths from the same start/goal configuration.

environment and agent information. By attaching annotations to an object in the environment, a user can provide useful positional information. These annotations are used to semantically construct customized prepositional constraints (for example, "in the grass," "near the wall," or "west of the train tracks"). Figure 6.1b illustrates some common annotations: Back, Front, Left, and Right, for a static object in the environment. The relationships between multiple objects can be similarly described by introducing annotations such as Between, shown in Figure 6.2c.

Dynamic objects can also easily be given spatial annotations. For example, we can denote an agent's LineOfSight; a simple approximation of this is shown in Figure 6.3. Note that it is conceptually simple (and supported by our framework) to calculate "true" line of sight, or other complex annotations, dynamically, as cost calculations can happen during path planner execution. In a world model which supports raycasting, true line of sight can be directly calculated in this way.

Annotations are defined independently from constraints and are used in the definition of a constraint to delimit its area of influence, based on objects, agents, and the environment. As such, annotations may be created ahead of time and freely reused by different constraint systems, which may run independently, allowing for rapid authoring of new constraint systems for new situations.

Hard Constraints

A hard constraint comprises just one field: an Annotation. This annotation represents an area in which states in Σ are pruned. Hard constraints can only be Not constraints; in order to specify hard attracting constraints, we use a sequence of goals that the agent must navigate to (e.g., go to the mailbox and then come back to the door). Hard constraints prevent all included transitions from being expanded during the search (by giving these transitions an infinite weight), thus producing a plan that always avoids the region. In addition, hard constraints can be used to model dynamic obstacles (any annotation that completely blocks off its contained states).

Soft Constraints

A soft constraint specification consists of three fields: (1) a preposition, (2) an annotation, and (3) the constraint weight. As soft constraints are the general case encompassing hard constraints (which are Not In with $w = -\infty$), they are simply referred to as "constraints" henceforth.

Preposition

We define two simple prepositions, Near and In, which define the boundaries of the region of influence. For example, we might wish to navigate Near a building (a fuzzily defined area of effect), while making sure that we are not In the grass (a well-defined area of effect). These two prepositions gain significant power by leveraging annotations placed in meaningful parts of the environment.

Weight

The weight defines the influence of a constraint, and can be positive or negative. For example, one constraint may be a weak preference ($w = 1$), while another may be a very strong aversion ($w = -5$); a negative weight indicates a repelling factor. Weights allow us to define the influence of constraints relative to one another (where one constraint may outweigh another), facilitating the superposition of multiple constraints in the same area with consistent results.

6.3.3   MULTIPLIER FIELD

Constraints must modify the costs of transitions in the search graph in order to have an effect on the resulting path generated. To achieve this, it is important to maintain several properties.

1. *The modified cost of a transition must never be negative, to ensure that the search technique will complete.* In our system, the cost of a transition will always be greater than or equal to its unmodified distance cost, even under the influence of attractor constraints. With A* and variants such as AD*, this guarantees optimality (and can prevent infinite loops and other instabilities) when the unmodified distance cost is used as a heuristic cost estimate.

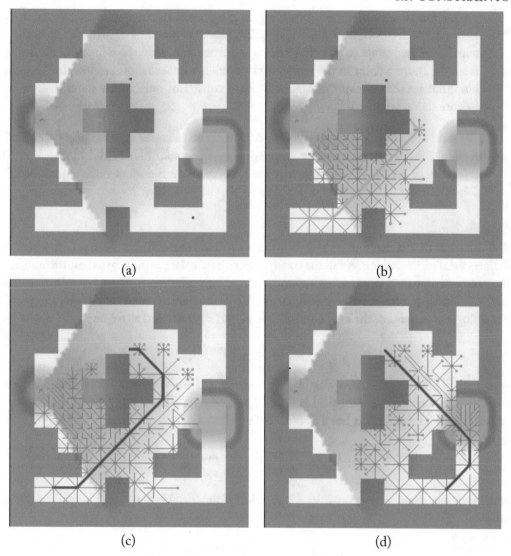

Figure 6.3: (a) Multiplier field incorporating a negative human relationship constraint [Sturtevant, 2013] and a positive area constraint. (b) "Partial" path planning result before planner has completed execution. (c), (d) Example plans showing the "Adaptive Highway" planning domain in this particular planning problem.

2. *We must be able to efficiently compute the cost of a transition, influenced by several constraints.* The weighted influence of all constraints at a particular state are simply summed together, and added onto the base weight, multiplying the cost.

3. *Constraints should only affect a limited region of influence.* In our system, soft constraints may have a smooth gradient or a hard edge. We model constraints as artificial potential fields and define the influence of a constraint at a particular position as a function of its distance from the nearest point in the annotation volume. A linear falloff ensures that a constraint has influence only at short distances. (A hard constraint only affects states that are within its annotation volume.)

4. *The total cost along a path must be independent of the number of states along that path.* To maintain this property, the cost calculation must be continuous, and modeled as a path integral (see below). The path integral will always have the same value regardless of path sub-division.

Formulation

The influence of a constraint is defined using a continuous multiplier field $m(\vec{x})$, where $m(\vec{x})$ denotes the multiplicative effect of the constraint at a particular position $\vec{x}$ in the environment. It is important to note that, due to its continuous nature, the multiplier field can be easily translated to any pathfinding system; it is not specific to graph search representations of pathfinding problems. For a single constraint c, the cost multiplier field $m_c(\vec{x})$ is defined as follows:

$$m_c(\vec{x}) = 1.1^{-W_c(\vec{x})},$$

where $W_c(\vec{x})$ is the constraint weight field (below) and the constant 1.1 is chosen for convenience[1]. The constraint weight field is defined as a position-dependent weight value for a constraint. For In constraints, it has a discrete definition:

$$W_c(\vec{x}) = \begin{cases} w & : & \vec{x} \in \text{annotation}_c, \\ 0 & : & \text{otherwise,} \end{cases}$$

while for Near constraints it provides a soft falloff with a fixed radius of $|w|$ outside of the annotation:

$$W_c(\vec{x}) = w \cdot \max\left(0, \frac{|w| - r_c(\vec{x})}{|w|}\right),$$

where $r_c(\vec{x})$ is the distance between the position $\vec{x}$ and nearest point in the volume of the annotation on the constraint c. Outside of the fixed radius $|w|$, a Near constraint has no effect.

This is especially important for dynamic constraints, as we must monitor all the states whose costs are updated, while performing plan repair. Explicitly defining the boundary of a constraint limits the number of states that a planner must consider for repair. Multiplier fields for Near attractor and repeller are visualized in Figure 6.6.

---

[1]It could be any value greater than one; its value only affects the range of useful w values.

Multiple Constraints

For a set of constraints $\mathbf{C}$, we define the aggregate cost multiplier field:

$$m_{\mathbf{C}}(\vec{x}) = \max\left(1, m_0 \prod_{c \in \mathbf{C}} m_c(\vec{x})\right) = \max\left(1, 1.1^{W_0 - \sum_{c \in \mathbf{C}} W_c(\vec{x})}\right).$$

To accommodate attractor constraints, which reduce cost, we define a "base" multiplier m_0 or base weight W_0, which is automatically calculated based on the weight values of the constraints in $\mathbf{C}$. This multiplier affects costs even in the absence of constraints, which allows attractors to reduce the cost of a transition while remaining above the original (Euclidean distance) cost. The resulting cost multiplier is thus limited to be ≥ 1, preserving optimality guarantees of the planner.

Cost multiplier for a transition

The cost multiplier for a transition $(s \to s_n)$, given a set of constraints $\mathbf{C}$, is defined as follows:

$$M_{\mathbf{C}}(s, s_n) = \int_{s \to s_n} m_{\mathbf{C}}(\vec{x}) \, d\vec{x}.$$

We choose to define this as a path integral because it is generalized to any path, not just a single discrete transition, and because it perfectly preserves cost under any path subdivision. For our graph representation, we estimate the path integral using a four-part Riemann approximation by taking the value of the multiplier field at several points along the transition.

6.4   PLANNING ALGORITHM

We use Anytime Dynamic A* (Chapter 5.3) as our underlying planner to efficiently repair solutions after world changes and agent movement. It quickly generates an initial suboptimal plan, bounded by an initial inflation factor ϵ_0 which focuses search efforts toward the goal. This initial plan is then improved by lowering the weight of ϵ gradually (while planning) until ϵ reaches 1.0, guaranteeing optimality of the final solution. AD* allows an anytime, dynamic approach to this weighted-region problem (previously solved using various other methods, see [Mitchell and Papadimitriou, 1991, Narayanappa et al., 2005, Reif and Sun, 2000]). The suboptimal solutions provided by the AD* inflation methods are used to find approximate solutions in real-time.

6.4.1   COST COMPUTATION

The modified cost of reaching a state s from s_{start}, under the influence of constraints, is computed as follows:

$$g(s_{\text{start}}, s) = g(s_{\text{start}}, s_n) + M_{\mathbf{C}}(s, s_n) \cdot c(s, s_n),$$

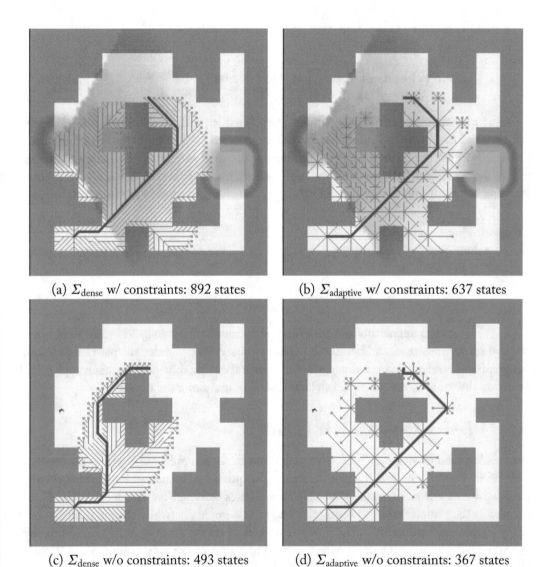

(a) Σ_{dense} w/ constraints: 892 states

(b) Σ_{adaptive} w/ constraints: 637 states

(c) Σ_{dense} w/o constraints: 493 states

(d) Σ_{adaptive} w/o constraints: 367 states

Figure 6.4: Graph expansion comparison between Dense Uniform and Adaptive Highway domains. (a) Dense Uniform navigation domain w/ constraints. (b) Adaptive Highway navigation domain w/ constraints. (c) Dense Uniform navigation domain, no constraints. (d) Adaptive Highway navigation domain, no constraints. Note in figures (c) and (d) that the resulting path is slightly different: this is due to the fact that the Adaptive Highway domain represents the environment with slightly less accuracy.

where $c(s, s_n)$ is the cost of a transition from $s \rightarrow s_n$, and $M_\mathbf{C}(s, s_n)$ is the aggregate influence of all constraint multiplier fields, as described in Section 6.3.3. This is recursively expanded to produce:

$$g(s_{\text{start}}, s) = \sum_{(s_i, s_j) \in \Pi(s_{\text{start}}, s)} M_\mathbf{C}(s_i, s_j) \cdot c(s_i, s_j),$$

which utilizes the constraint-aware multiplier field to compute the modified least-cost path from s_{start} to s, under the influence of active constraints $\mathbf{C}$. Each state can keep track of its set of influencing constraints to mitigate the need to exhaustively evaluate every constraint repeatedly. When the area of influence of a constraint changes, the states are efficiently updated, as described below.

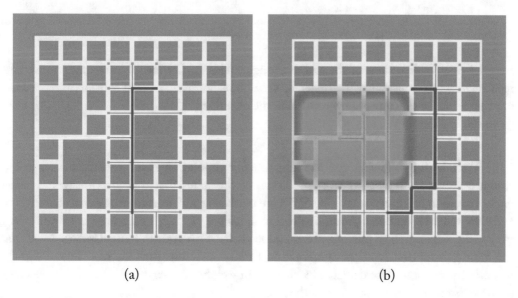

(a)   (b)

Figure 6.5: Demonstration of our framework applied to a map navigation problem using examples in a city grid layout. (a) A simple no-constraint solution. (b) Addition of a `Not Near NeighborhoodX` constraint, on the same map. For example, the negative constraint might be used at night to avoid a dangerous neighborhood, but during the day that constraint may be removed or weight-reduced.

6.4.2   ACCOMMODATING DYNAMIC CONSTRAINTS

Over time, objects associated with a constraint may change in location, affecting the constraint multiplier field, influencing the search. For example, an agent constrained by a `LineOfSight` constraint may change position, requiring the planner to update the plan to ensure that the constraint is satisfied. Each constraint multiplier field $m_c(\vec{x})$ has a region of influence $\mathbf{region}(m_c, \vec{x})$, which defines the finite set of states $\mathbf{S}_c$ that is currently under its influence.

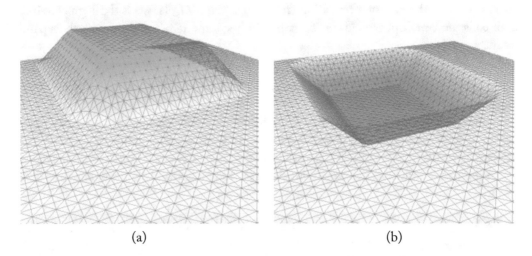

(a)   (b)

Figure 6.6: 3D depictions of the additive weight fields $W_c(\vec{x})$ for (a) attracting/positive and (b) repelling/negative constraints. These demonstrate the falloff properties of soft Near constraints and provide an intuitive picture of how constraints are quantified.

We take the following approach for moving annotation-based constraints: when a constraint c moves from $\vec{x}_{\text{prev}}$ to $\vec{x}_{\text{next}}$, the union of the states that were previously and currently under its region of influence ($\mathbf{S}_c^{\text{prev}} \cup \mathbf{S}_c^{\text{next}}$) are marked as inconsistent (their costs have changed) and they must be updated. Additionally, for states $s \in \mathbf{S}_c^{\text{next}}$, if c is a hard constraint, its cost $g(s) = \infty$. Algorithm 16 provides the pseudocode for **ConstraintChangeUpdate**. Note that, in the more general case (with non-annotation-based constraints such as "real" line of sight), the inconsistent region can be detected by changes in weights at every node that may have changed (or other implementation-specific methods).

Finally, if the navigation graph has changed, the routine **UpdateState**(s) is used to recompute the costs of states. Another possible approach would be to periodically update the plan instead of detecting when it is necessary. Our approach has an advantage in more dynamic environments, where periodic re-planning may be too infrequent. In the vehicle example (Figure 6.12), less frequent updates would result in collisions with the vehicle hazards.

In actual execution, we note that re-planning time can sometimes be considerable. To maintain real-time performance without continuing along the old, incorrect trajectory, the system will report no known solution until a partial or full plan is known (depending on the exact algorithm conventions). As a result, a "stopping"/"waiting" behavior is observed: an agent may pause its motion while recalculating to find the best way to continue. In most cases, this stopping time will be short. With algorithms able to provide suboptimal paths (such as AD*), the agent may follow these solutions while recalculating.

Algorithm 16 - Update due to change in constraints

Input: Constraint c, previous position $\vec{x}_{\text{prev}}$, next position, $\vec{x}_{\text{next}}$

1: **ConstraintChangeUpdate** $(c, \vec{x}_{\text{prev}}, \vec{x}_{\text{next}})$;
2: $\mathbf{S}_c^{\text{prev}} = \mathbf{region}(m_c, \vec{x}_{\text{prev}})$;
3: $\mathbf{S}_c^{\text{next}} = \mathbf{region}(m_c, \vec{x}_{\text{next}})$;
4: **for each** $s \in \mathbf{S}_c^{\text{prev}} \cup \mathbf{S}_c^{\text{next}}$ **do**
5:     **if** $(\text{pred}(s) \bigcap \text{VISITED} \neq \text{NULL})$ **then**
6:         **UpdateState**(s);
7:     **end if**
8:     **if** $(s_n \in \mathbf{S}_c^{\text{next}} \wedge c \in \mathbf{C}_h)$ **then**
9:         $g(s_n) = \infty$;
10:       **if** $(s_n \in \text{CLOSED})$ **then**
11:         **for each** $s'' \in \text{succ}(s_n)$ **do**
12:           **if** $(s'' \in \text{VISITED})$ **then**
13:             **UpdateState**(s'');
14:           **end if**
15:         **end for**
16:       **end if**
17:     **end if**
18: **end for**

6.5 RESULTS

Our initial code base is implemented in C# and uses the Unity game engine. The ADAPT platform [Shoulson et al., 2013b] was used for character animation where applicable. Our revised code base is implemented as a stand-alone .NET library in C# along with a layer for integrating with Unity. This revised code includes implementations of the various planners (A*, ARA*, AD*, etc.) and the system for handling stating and dynamic constraints. It is published at the following URL and is also available by request to the authors:

`https://bitbucket.org/kainino/constraint-aware-navigation`

Our framework meets strict time guarantees by publishing sub-optimal paths within time constraints (e.g., within the span of one frame) and iteratively refines the plan in subsequent frames while interleaving path planning and plan execution. In practice, our framework is able to converge to an optimal path within a few frames, and additional planning time is only needed to handle dynamic constraints and goal changes. For large changes that invalidate the current path, the AD* algorithm repairs the solution efficiently by increasing the inflation factor (temporarily trading optimality for computational efficiency) and refining the plan as soon as time permits.

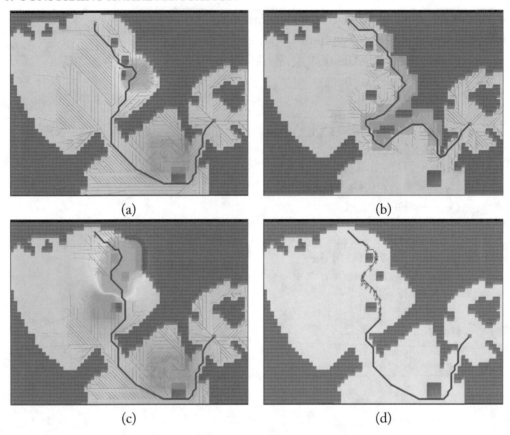

Figure 6.7:   Navigation under different constraint specifications. (a) Attractor to go behind an obstacle and a repeller to avoid going in front of an obstacle. (b) Combination of attractors to go along a wall. (c) Combination of attractors and repellers to alternate going in front of and behind obstacles, producing a zig-zag trajectory. (d) Emulating lane formation using multiple agents under the same constraints.

6.5.1   EVALUATION OF HYBRID DOMAIN

The node expansion count (correlated with the cost-depth of the search) dominates the computational complexity of the search. The use of *highway* transitions (transitions from $\mathbf{A}_{tri}$) reduce the search depth, as the length of a utilized transition from $\mathbf{A}_{tri}$ is, on average, much longer than a transition in $\mathbf{A}_{dense}$ (depending on the triangulation method and the grid density, respectively; in our examples, around 2–6 times longer). Figure 6.8 compares the use of Σ_{hybrid} and Σ_{dense} for the same problem instance. We observe that there is a reduction from 145 to 90 nodes expanded, for just 4 highway nodes used in the plan for the small problem instance in Figures 6.8a–b. The problem instance in Figure 6.8c–d is particularly challenging for the planner as the heuristic focuses the search in directions that are ultimately blocked. This leads to a significantly greater exploration

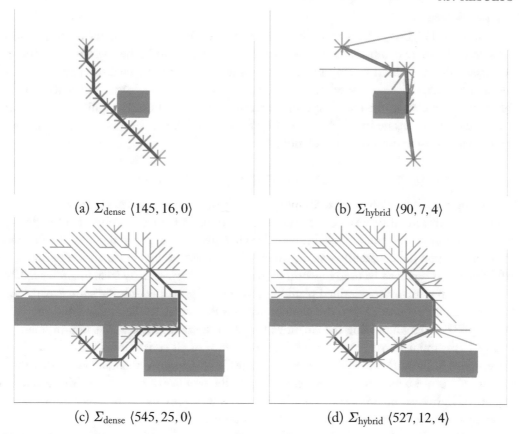

(a) Σ_{dense} $\langle 145, 16, 0 \rangle$

(b) Σ_{hybrid} $\langle 90, 7, 4 \rangle$

(c) Σ_{dense} $\langle 545, 25, 0 \rangle$

(d) Σ_{hybrid} $\langle 527, 12, 4 \rangle$

Figure 6.8: Comparative evaluation of dense and hybrid domains. Blue indicates transitions in $\mathbf{A}_{\text{dense}}$, red indicates highway transitions from $\mathbf{A}_{\text{tri}}$. Numbers shown are $\langle$number of nodes expanded, number of dense nodes chosen in path, number of highway nodes chosen in path$\rangle$. (Some minor issues with the connectivity of the graph, due to nearby obstacles, cause some localized suboptimalities in the path.)

of nodes in Σ_{dense} before a solution can be found, and dilutes the benefits of highway selection: We see a reduction from 545 to 527 nodes here.

Based on our experiments, we observe that the number of highway nodes n_h used in the final plan reduces the number of nodes expanded in the search by around $10n_h$ nodes in small environments. During suboptimal planning (with an inflation factor greater than one), well-aligned highway transitions can help guide the planner to a suboptimal solution. This varies depending upon the environment configuration, the number and type of constraints used, and where in the plan a highway node is chosen. (The earlier a highway node is chosen during plan computation, the more significant its impact on the reduction in node expansion.)

Highway Selection

The selection of highway nodes depends on the quality of triangulation, and the relative position of the start and goal, in comparison to where these nodes are present in the environment. This could be potentially mitigated by using navigation meshes with different qualities [Kallmann, 2014, Mononen, 2015, Oliva and Pelechano, 2011, 2013, Pettré et al., 2005, van Toll et al., 2012]. The inflation factor used in the search also influences highway selection. For a high inflation factor, the search is more prone to selecting highway nodes since they accelerate and focus the search while compromising optimality of solution.

6.5.2   EVALUATION OF ADAPTIVE HIGHWAY DOMAIN

The Adaptive Highway domain works independently of any triangulated representation or other highly subjective input, instead lazily pruning transitions to reduce the number of node expansions and improve computational efficiency. This has advantages due to the close dependence of the domain upon the constraint system, which allows more intelligent expansion optimizations. Additionally, it can even be used in world models with no high-quality triangulation available.

Even in a highly constrained planning problem (Figure 6.4a–b), the state expansions are still significantly reduced. Here, the planner expands 892 states in the Dense Uniform domain, but only 637 states when using the Adaptive Highway domain; this is a 30% improvement, even in a very constraint-dense environment. In the absence of all constraints (Figure 6.4c–d), state expansion is improved from 493 states to 367 states, a 25% improvement, showing that the performance improvement is not very dependent upon the number of constraints. This is due to the Adaptive Highway domain's awareness of the constraint system: its performance improvement is generally more significant in the presence of constraints.

Figures 6.9, 6.10, and 6.11 show plots of planner results over the amount of time spent planning. This is independent of the amount of time given to the planner to plan each frame.[2] Planning time measurements are summarized in Table 6.1.

Table 6.1: Planning time measurements (total sum of program time spent planning) with various domains and planners

Plan time/Plan cost	AD*, Dense	AD*, Adaptive	A*, Adaptive
To suboptimal plan	0.84s / 76	0.22s / 83	—
To optimal plan	1.94s / 68	0.50s / 73	0.19s
Plan repair	0.33s / 65	0.04s / 69	0.18s

This data shows a significant performance improvement when using the Adaptive Highway domain, despite a small reduction in path optimality. Although total planning time for A* is better

---

[2]Depending on the importance of the plan results, more or less time can be given to the planner to plan; for example, an autonomous robotic agent might dedicate a full CPU core to planning, while a mobile game might try to minimize time spent planning.

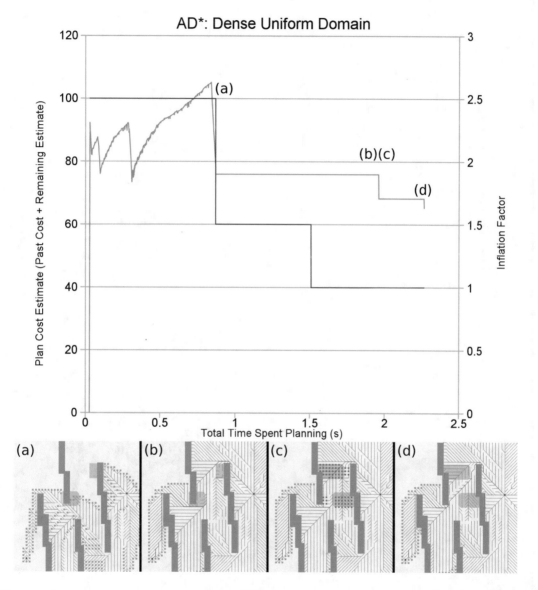

Figure 6.9: Plan cost vs. time plot for Dense Uniform domain. Planning is rather time-intensive but very accurate. Partial solutions are provided until (a) a suboptimal solution is found. (b) An optimal solution is found. (c) The constraints mutate, producing "invalid" states. (d) A new optimal solution is found.

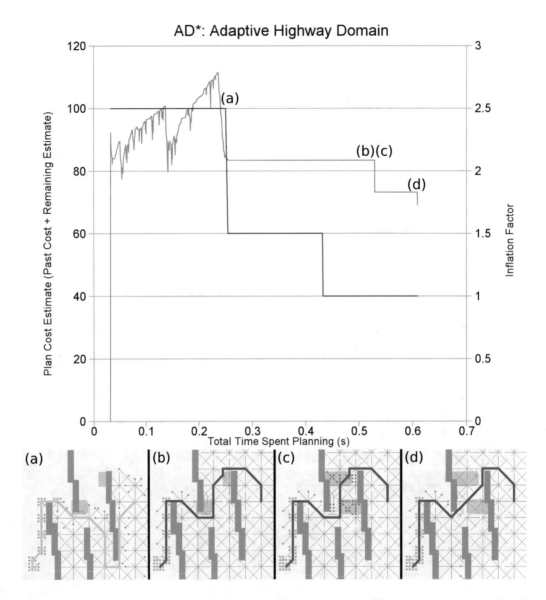

Figure 6.10: Plan cost vs. time plot for Adaptive Highway domain. Total planning time is found to be much shorter than for the Dense Uniform domain, at the cost of a small amount of plan accuracy. Partial solutions are provided until (a) a suboptimal solution is found. (b) An optimal solution is found. (c) The constraints mutate, producing "invalid" states. (d) A new optimal solution is found.

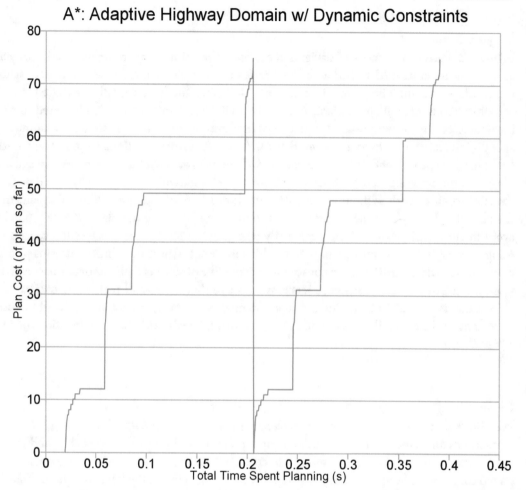

Figure 6.11: Plan cost vs. time plot for A* in the Adaptive Highway domain. The same scenario is shown, without dynamic plan repairs. Instead, A* is run again from scratch after the constraints are modified.

than for AD*, note that the relative planning time for A* is much longer than for AD*, so re-planning with A* is still impractically slow, especially in larger environments.

6.5.3   CONSTRAINT SYSTEM EXAMPLES

Simple examples

Figure 6.2 illustrates a variety of navigation examples for a simple environment. Static obstacles and agents are annotated to add additional nodes in the triangulation to accommodate spatial relationships including Between, Front, Back, Left, etc. The hybrid graph, illustrated in 6.2b, combines the transitions in Σ_{tri} and Σ_{dense}. A specific problem instance **P**, illustrated in 6.2c, includes a start, goal configuration, and a set of hard and/or soft constraints. In this example, the agent is instructed to go Near Between B and C (a soft attractor), Not Near LineOfSight of the agent (a soft repeller), and Not In the grass (a well-defined area with a soft repeller). Image 6.2e illustrates the resulting node expansion and path produced, which is drastically different from the static optimal path without any constraints, shown in 6.2d. Images 6.2f and 6.2g illustrate the efficient plan repair to accommodate constraint changes where the plan must be refined to avoid the line of sight of a moving agent. By changing the relative influence of the constraints using constraint weights w, we can produce different results where one constraint gains priority over another. In this example, the constraint to avoid line of sight is stronger than the constraint to stay between the two obstacles. Hence, we observe that if no valid path exists that satisfies all constraints, a solution is produced that accommodates as many constraints as possible, based on weights. Image 6.2h illustrates multiple agents planning with different combinations of constraints.

Game environments

We also demonstrate the method on challenging game environments [Sturtevant, 2012]. The method of constraint specification using simple prepositional phrases is extensible, and simple atomic constraints can be easily combined to create more complex, composite constraints. Compound constraints like staying along the wall or alternating between the left and right of obstacles to produce a zig-zag path can be created by using combinations of multiple attractors and repellers, as shown in Figures 6.7a–c. Figure 6.7d illustrates multiple agents conforming to a common set of constraints in their paths, emulating a lane-formation or single-file behavior. (However, our method is *not* suitable to crowd simulation, as there is a high computational and memory cost to run one instance of the planner. It is intended for use with a small number of agents which must navigate intelligently.)

Figure 6.12 shows the use of constraints in a road crossing scenario, where the agent avoids navigating in front of moving vehicles. While this demo does not show a very practical use of constraint planning, it illustrates considerable robustness in such a rapidly changing environment even without any awareness of obstacle trajectory.

Plan repair to avoid the line of sight of multiple moving agents is shown in Figure 6.13. Here, the user interactively selects agents associated with the constraints and changes their position, thus invalidating the current plan. The same problem configuration using attractor con-

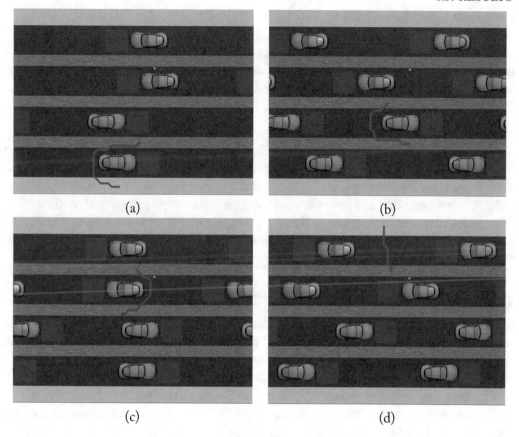

Figure 6.12: Navigation under different constraint specifications: dynamic constraints used to avoid navigating in front of vehicles as a proof of robustness.

straints for `LineOfSight` produces a drastic change in the resulting path, as shown in Figure 6.14. Our framework efficiently repairs the existing solution to accommodate the constraint changes.

Human relationship constraints

The work of Sturtevant [2013] incorporates human relationships (personal distance and line of sight) into path planning using a similar approach to our framework. In [Kapadia et al., 2013d], we used a rough approximation of line of sight to demonstrate dynamic constraints. In the updated framework, we use a more robust human relationship constraint modeled after Sturtevant's human relationship constraints. Figure 6.3 shows a simple environment which uses this constraint.

Large-scale environments

While most of our examples operate at human movement scales, our framework can operate at other scales, such as map navigation. This can be useful not only for autonomous agent planning, but also for automatic route planning. In Figure 6.5, a simple neighborhood-avoidance problem is demonstrated which operates in a grid-based city layout.

Spatiotemporal constraint specification

Another type of problem we have briefly explored, to illustrate support for other useful tactical behavior specifications, is that of time- and date-based constraints. Since the framework allows for fully dynamic constraints, it is possible to create spatial constraints which change weight (or disappear) before or during execution, depending on some varying input such as time of day or other agent/environment status (e.g., "Are the guards alerted? If so, avoid some regions."). In Figure 6.5, we show a planning problem with varying weight in order to avoid a certain neighborhood at night.

6.5.4 PARAMETER SELECTION AND PERFORMANCE

For our experiments, ϵ was initially set to a value of 2.5 to quickly produce a sub-optimal solution while meeting time constraints, which could be iteratively refined over subsequent plan iterations. t_{max} was set to 0.032 s and the plan computations of multiple agents were distributed over successive frames, to ensure that the frame rate was always greater than 30 Hz. The maximum allotted time can be further calibrated to introduce limits on computational resources or accommodate many characters, at the expense of plan quality. We observed that in general, the value of ϵ quickly converges to 1.0 to produce an optimal path, and requires only a few frames to repair solutions to accommodate dynamic events (see Table 6.1). For rapid changes in the environment over many frames, the planner may be unable to find a solution and the agent stops moving till a valid path is computed for execution.

The AD* algorithm requires all visited nodes in the search graph to be cached in order to facilitate efficient plan repair, imposing a memory overhead for large environments. There exists a trade-off between computational performance and memory requirements where using a traditional A* search would require less nodes to be stored, at the expense of planning from scratch whenever the plan is invalidated.

The choice of the base multiplier m_0 impacts how constraints affect the resulting cost formulation, with higher values diluting the influence of the distance cost and the heuristic on the resulting search. We automatically pick the lowest possible value of m_0 to accommodate the maximum value of attractor constraints while preserving optimality guarantees. A cost model where the base multiplier has no adverse effect on admissibility or the influence of the heuristic is the subject of future work.

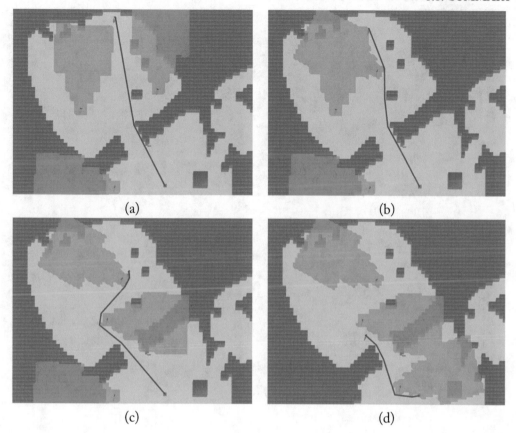

(a)   (b)

(c)   (d)

Figure 6.13: `Not In LineOfSight` constraint. Utilizes multiple dynamic agents. A user interactively moves agents and the plan is repaired to accommodate constraint change.

6.6   SUMMARY

One problem that is usually overlooked in planning research is to impose arbitrary spatial and temporal constraints on the paths that are generated. To meet this important requirement we have presented a goal-directed navigation system that satisfies multiple spatial constraints imposed on the paths. Constraints can be specified with respect to obstacles in the environment, as well as other agents. For example, a path to a target could be altered to stay behind buildings and walk along walls, while avoiding line of sight with patrolling guards. An extended anytime-dynamic planner is used to compute constraint-aware paths, while efficiently repairing solutions to account for dynamic constraints.

The hybrid environment representation described in Chapter 6 is sensitive to the kind of triangulations produced for the environment. The work can be easily integrated with more sophisticated triangulation strategies such as those described in Chapter 4, and with manually annotated

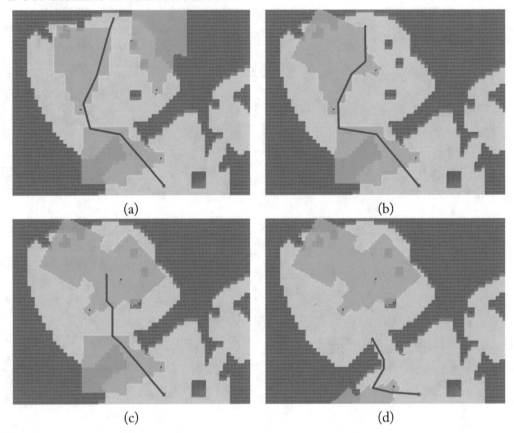

Figure 6.14: In `LineOfSight` constraint. Utilizes multiple dynamic agents. A user interactively moves agents and the plan is repaired to accommodate constraint change.

waypoint graphs in order to improve computational performance. The performance improvements of the Adaptive Highway domain are contingent upon the constraint weights and their area of influence, which need to be addressed to provide an efficient discrete environment representation that generalizes across different constraints.

Static analysis of the environment could potentially yield automatic annotation generation (e.g., `Between`, `Inside`), which would improve creating spatial relationships, and thus is an interesting subject of future exploration. Additionally, our ahead-of-time planning could be extended to include trajectory extrapolation by using modified annotation positions based on search depth; this could improve plans in environments like the road-crossing example. Finally, we have most closely considered spatial constraints in this chapter, but our framework is general and extensible to other problem domains; we would like to expand upon other types of planning problems. Although we briefly explored spatiotemporal constraints, we see many possible ways to expand this to more advanced problems.

To highlight the benefits of our method, all constraints were accounted for at the global planning layer. However, in some cases where the constraint is constantly changing, such as a moving vehicle, it may be significantly more efficient to use a locally optimal strategy for constraint satisfaction. A hybrid approach that combines the benefits of both global planning and local collision-avoidance for constraint satisfaction is another subject of future exploration.

CHAPTER 7

Anytime Dynamic Search on the GPU

Thus far we have systematically presented foundational approaches for planning in complex domains, and extensions to handle dynamic environments while meeting real-time constraints. While researchers have explored several methods to make real-time planning a more tractable problem, and many of these techniques have been very successful, i.e., dynamic planners can efficiently adapt to dynamic changes in a graph, hierarchical planners increase performance by breaking down the problem at different levels of abstraction. However, GPU planning remains a widely unexplored area of research.

In this chapter, we advocate that the GPU is a tool that can be leveraged for much of the computational burden for planning and that it should be used when appropriate. We present a parallel wavefront based anytime planning approach that can harness the power of the GPU to provide orders of magnitude speedup for multi-agent pathfinding applications. We improve on our method by partitioning the state space using quad-trees and performing adjacency detection via quad-codes. Our proposed algorithms harness the power of graphics hardware for real-time mult-agent path planning in large, complex, dynamically changing environments. Where other approaches rely on local collision avoidance to meet real-time constraints, our results demonstrate the potential for global planning operations for tens of hundreds of agents at interactive rates. For additional details, please refer to [Garcia et al., 2014, Kapadia et al., 2013c].

7.1 GPU PLANNING ON UNIFORM GRIDS

We first introduce a parallelizable, wavefront-based approach to dynamic path planning. This technique exploits graphics hardware to considerably reduce the computational load, while still maintaining strict optimality guarantees. In addition, this approach is able to perform efficient updates to accomodate world changes and agent movements, while reusing previous computations.

We analyze different termination conditions and introduce a strategy which preserves optimality guarantees, even on search graphs with non-uniform costs, while requiring minimum GPU iterations. Furthermore, the computational complexity of this approach is independent of the number of agents, facilitating optimal, dynamic path planning for a large number of agents in complex dynamic environments, opening the possibility to large-scale crowd applications. In particular, our proposed solution offers the following main contributions.

- A wave-front based search technique that can efficiently handle world changes and agent movement, while reusing previous efforts, and is amenable to massive parallelization.

- A termination condition which enforces strict optimality guarantees, even for non-uniform search graphs, while requiring minimum number of GPU iterations.

- Extension to handle any number of moving agents, at no additional computational cost.

7.1.1 METHOD OVERVIEW

Our method relies on appropriate data transfer between the CPU and GPU at specific times. Given a state space of size $n \times m$, $n \times m$ states are allocated in the GPU to represent the entire environment by calling $generateMap(n, m)$. Every unoccupied state s_u is initialized to $g(s_u) = -1$ and every obstacle state s_o is initialized to $g(s_o) = \infty$, where $g(s)$ represents the current g-value for state s. Since each expanded state updates its current g-value, we can assume that $g(s) = -1$ means that state s has not been updated. We also assume that if $g(s) = \infty$ then s is an obstacle.

Given an environment configuration with start and goal states, s_s and s_g, correspondingly, the planner invokes a GPU kernel repeatedly until a solution is found. To do so, two copies of the state space, S_r and S_w, are created. S_r is used to read the current state g-values while S_w is used to write the updated values for the new iteration. After each iteration (e.g., kernel execution), S_r and S_w are swapped such that in the next iteration S_r becomes S_w and vice-versa. This strategy addresses the synchronization issues inherent in GPU programming. Once the planner is done executing, a solution can be found by following the least cost path from s_g to s_s.

To adapt to the dynamic setting, once a plan is generated, the algorithm waits for changes in the environment. If the system observes an obstacle movement from state s to state s', S_r and S_w are updated by setting $g(s') = \infty$ and $g(s) = -1$. s' is now marked an obstacle and s needs to be updated. In addition, every neighbor s_n of s', where $s_n \neq s$, is as inconsistent if s' is its least cost predecessor. The planner kernel monitors states that are marked as inconsistent and efficiently computes their updated costs (while also propagating inconsistency) without the need for replanning from scratch.

Change in start state can also be efficiently handled by performing the search in a backward fashion from the goal to the start, and marking the start previous state as inconsistent to ensure a plan update. Algorithm 17 provides the pseudocode for the plan solver *computePlan*. This function is executed in the CPU.

7.1.2 GPU-BASED WAVEFRONT ALGORITHM

Existing graph search methods [Hart et al., 1968] guarantee optimality and work well for dynamic environments [Likhachev et al., 2005]. However, they are not amenable to massive parallelization. The wavefront algorithm [Pal et al., 2011] takes its name as an analogy for the way it behaves. It sets up the environment with a initial state which contains some initial cost. At each iteration,

Algorithm 17 *computePlan*(S_{cpu})

$S_r \leftarrow S_{cpu}$;
$S_w \leftarrow S_{cpu}$;
repeat
    $flag \leftarrow 0$;
    plannerKernel(S_r, S_w, $flag$);
    swap (S_r,S_w);
    $incons \leftarrow false$;
    for each s **in** S_r **do**
        if ($incons(s) = true$) **then**
            $incons \leftarrow true$;
            break;
        end if
    end for
until ($flag = 0 \wedge incons = false$)
$S_{cpu} \leftarrow S_r$;

---

every state at the frontier is expanded computing its cost relative to its predecessors' cost. This process repeats until the cost for every state is computed, thus creating the effect of a wave spreading across the state space. Wavefront based approaches are inherently parallelizable, but existing techniques require the entire map to be recomputed to handle dynamic world changes and agent movements. Figure 7.1 visualizes the wavefront propagation process in a simple environment.

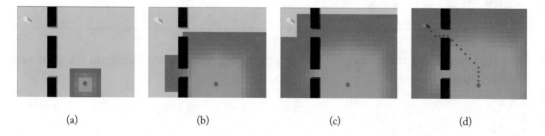

(a)   (b)   (c)   (d)

Figure 7.1: Wavefront expansion process: (a) 3 iterations; (b) 11 iterations; (c) 15 iterations; and (d) 18 iterations.

Algorithm 18 describes the shortest path wavefront algorithm ported to the GPU. The planner first initializes the cost of every traversable state to a default value, $g(s_u) = -1$, indicating it needs to be updated. States occupied by obstacles take a value of infinity, $g(s_o) = \infty$, and the goal state is initialized with a value of 0, $g(s_g) = 0$. The planner finds the g-value for reaching any state s_u from the s_g by launching a kernel at each iteration that computes $g(s)$ as follows:

$$g(s) = min_{s' \in succ(s) \wedge g(s') \geq 0}(c(s, s') + g(s')).$$

For each successor that has been updated (i.e., $g \geq 0$). This process continues until all states have been updated. As discussed before, to address the concurrency problem inherent in parallel applications, two copies of the state space are created S_w and S_r. The kernel launches $n \times m$ threads, where each thread t_s reads the data from S_r corresponding to state s, and updates s in S_w. This ensures that we read from memory that will not change as we are executing the kernel.

Once the kernel finishes execution, S_r and S_w are swapped, allowing the threads to read the most up-to-date values for every state.

Algorithm 18 *plannerKernel*($^*S_r$, $^*S_w$, $^*f$)

$\quad s \leftarrow state(t_s)$;
$\quad$**if** $(s \neq s_o \wedge s \neq s_g)$ **then**
$\quad\quad$**for each** s' in $neighbor(s)$ **do**
$\quad\quad\quad$**if** $(s' \neq s_o)$ **then**
$\quad\quad\quad\quad newg \leftarrow g(s') + c(s, s')$;
$\quad\quad\quad\quad$**if** $((newg < g(s) \vee g(s) = -1) \wedge g(s') > -1)$ **then**
$\quad\quad\quad\quad\quad pred(s) \leftarrow s'$;
$\quad\quad\quad\quad\quad g(s) \leftarrow newg$;
$\quad\quad\quad\quad\quad$\{ evaluate_termination_condition \}
$\quad\quad\quad\quad$**end if**
$\quad\quad\quad$**end if**
$\quad\quad$**end for**
$\quad$**end if**

The kernel also takes a flag, f, as a parameter which is set depending on the termination condition used:

Exit when start is reached
Before each kernel run the flag is set to $f = 1$. If in the current iteration $g(s_s) \geq 0$, that means there is a plan from s_g to s_s and the execution can be terminated. This is signaled by setting $f = 0$. While this approach will require considerably fewer iterations, it will not guarantee optimality on search graphs with non-uniform costs. The test is:

$$\text{if}(s == s_s)\text{f} = 0.$$

Exit on convergence

An alternate exit condition is to continue propagating until the whole map has been updated. This approach will guarantee optimality but with a considerable increase in the number of iterations. In this case, before each kernel run f is set to 0. If any update occurs at a given iteration, f is set to 1 ensuring that the planner keeps running until no further update is possible. In other words, the planner terminates only when the cost computation for the entire environment has converged. This will compute costs for unnecessary parts of the environment but will guarantee optimal solutions. Flag f is set as 1 as long as needed:

$$f = 1.$$

Minimal Map Convergence with Optimality Guarantees

The naive approach discussed previously does much more work than necessary to find an optimal plan. For large environments, this is extremely expensive. We introduce a termination condition that can greatly reduces the number of iterations required to find an optimal plan in large environments with non-uniform search graphs. If at any iteration, we find that the minimum g-value expanded corresponds to that of s_s, this means that a plan from s_g to s_s is available and any other possible path would yield a higher cost. To make sure that the s_s is expanded at each iteration (to compare to the other states expanded), $g(s_s)$ is set to -1 before each kernel run, marking it as a state that needs to be updated. To implement this strategy, it is enough to just adjust the condition that would set the flag that terminates the execution:

$$\textbf{if}(g(s) < g(s_s) \vee g(s_s) = -1)f = 1.$$

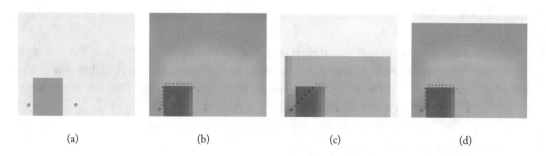

(a)	(b)	(c)	(d)

Figure 7.2: Comparison of termination conditions. (a) Non-uniform state space. The states shown in red are of much higher cost as compared to other states. (b) The planner terminates after updating g values for the whole map, producing an optimal path with significantly more iterations = 17. (c) Plan termiation as soon as it finds a path to the goal, producing a sub-optimal path. Total number of iterations = 8. (d) The proposed termination condition requires minimal iterations, while providing strict optimality guarantees. Number of iterations = 12.

7.1.3   EFFICIENT PLAN REPAIR FOR DYNAMIC ENVIRONMENTS AND MOVING AGENTS

In dynamic environments, changes in edge cost may invalidate plans while they are being executed. For example, and obstacle may obstruct a path, thus preventing running the plan to completion.

To handle dynamic changes we incorporate a flag that marks a state, s, as inconsistent if its predecessor, $pred(s)$, is not the neighbor with lowest cost or if its successor, $succ(s)$, is inconsistent. Suppose an obstacle moves from state s to state s', then $g(s') = \infty$ and $g(s) = -1$ (marking it for update). For each successor s'' of s', s'' is inconsistent if $s' = pred(s'')$, so $g(s'')$ is set to -1, forcing it to update. In other words, if any of the neighbor states has s' as a predecessor, they are flagged as inconsistent, thus mandating an update.

The main kernel then propagates in the same fashion until the termination condition is satisfied and no inconsistent states along the optimal path are left. By appending the following code to the end of the *plannerKernel* (Algorithm 19), it is guaranteed that node inconsistency will be propagated and resolved in the entire map.

Algorithm 19 Algorithm to propagate state inconsistency

$s \leftarrow state(t_s)$;
$incons(s) \leftarrow false$;
if $(incons(pred(s)) = true)$ **then**
  $incons(pred(s)) \leftarrow false$;
  $incons(s) \leftarrow true$;
  $g(s) \leftarrow -1$;
end if

Handling changes in s_s is straightforward. For the non optimized planner, the cost to reach every state has already been computed, so a path from s_g to s_s is always available. For the optimized exit condition, it is necessary to run the planner again to ensure that any state along the optimal path between s_g and s_s is expanded.

7.1.4   MULTI-AGENT PLANNING

For a multi-agent approach, we can define several start states, each corresponding to the current state of an agent. Our GPU planner is able to handle multiple agents and create paths for each of them in such a way that they avoid collisions with each other. We interleave planning and execution by running the kernel every time an agent moves. Each agent is able to update its own plan by following the least cost path from its position to the goal each time the planner kernel is run. We extend our planner implementation by making a slight modification to the termination condition to account for multiple agents. We execute the kernel until the agent with the largest

g has a smaller g that any of the non-agent states that were expanded for that iteration, and all agent states had a chance to be reached:

$$\mathbf{if}((g(s) < max_{a_i \in \{a\}} g(a_i)) \vee (g(a_i) = -1 \forall a_i \in \{a\})).$$

This means that the number of iterations it will take the planner to finish execution will depend on the distance from the goal to the farthest agent and not in the number of agents. Once the environment costs have been updated, each agent simply follows the least cost path from the goal to its position to find an optimal path.

Multi-Agent Simulation

It is simple to extend this planner to simulate a crowd of autonomous agents since it can efficiently adapt to environment changes even for a large number of agents. Grid states currently occupied by an agent incur an additional cost, which impacts the manner in which the wave propagates through the state space. Each agent is simulated to move along its current path using a simple particle simulator (the path is guaranteed to avoid obstacles and agents). At each frame, the map is repaired to accommodate world changes and agent movements, thereby obtaining valid plans for all agents.

Multiple Target Locations

This framework is currently limited to a small number of target locations, since it requires an entire copy of the state space to be maintained for each target, resulting in a significant memory overhead. This approach is very limited when dealing with multiple agents each with its own goal. In this situation, there would need to be a copy of the entire environment for each agent, requiring a great amount of memory (a scarce resource on the GPU). This issue is addressed in Section 7.2 by using an adaptive environment representation based on quad-trees.

7.1.5 RESULTS

This planner was tested on several challenging navigation benchmarks [Sturtevant, 2012] to showcase the benefits and limits over traditional methods using two different GPUs. Table 7.1 gives the specification of both units.

Figure 7.3 demonstrates the scalability of our approach as the number of agents is increased on a 256×256 environment. We observe that there was no noticeable increase in the computational cost with increase in number of agents.

Figure 7.4 illustrates the ability of this framework to accommodate to large environments. We tested it with a single agent with a goal distance of $N/2$ in a $N \times N$ world map. We observe that the use of the minimal yet sufficient exit condition (EXIT A) produces significant performance improvements over EXIT B as the planner does not have to wait until the g-values of the

Table 7.1: Graphics processing units (GPU) specifications

Information	GPU 1	GPU 2
Type	Geforce GT 650M 2GB	GeForce GTX680
Warp Size	32	32
Threads/Block	1024	1024
Global Mem	2,147,483,648 Bytes	2,147,483,648 Bytes
MultiProcessors	2	8
Mem Clock Rate	900,000 KHz	3,004,000 KHz
Mem Bus Width	128 bits	256 bits
Chip Clock Rate	950,000 KHz	1,058,500 KHz

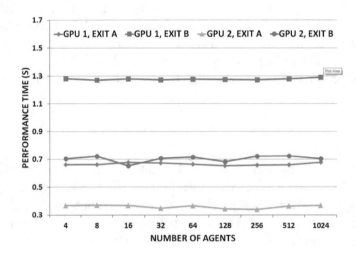

Figure 7.3: GPU planner performance with increase in number of agents. EXIT A: Exit condition that checks for convergence of only agent states. EXIT B: Exit conndition which checks for convergence of whole map. All solutions returned are optimal paths. Experiment performed on a 256 × 256 environment. GPU memory = 5120 KB. CPU memory varies from 2048–2080 KB.

entire state space have converged. Figure 7.5 illustrates the performance of our approach on a variety of challenging benchmarks [Sturtevant, 2012].

Figure 7.6 illustrates the overall advantage of using our method with a test scenario handling obstacle, agent and goal movement. We generated a random map of size 512 × 512 populated with 8 agents. The graph shows that our method took fewer iterations to reach an optimal solution at each step, with a significant performance improvement on the initial plan and after goal movement, when a solution needs to be found from scratch.

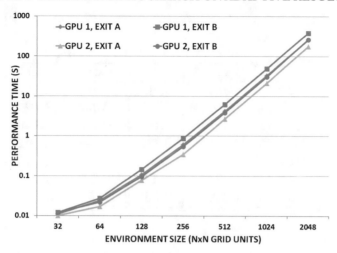

Figure 7.4: The figure shows the running time required to find an optimal solution different environment sizes on the average case. The benefits of this approach are more noticeable on larger state spaces.

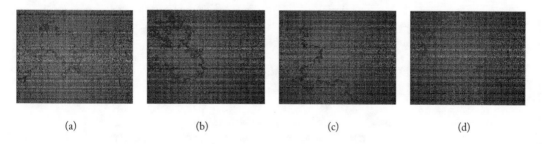

Figure 7.5: Global navigation for multiple agents on a variety of challenging benchmarks [Sturtevant, 2012] of size 512 × 512.

Figure 7.7 illustrates the memory requirements of our approach based on world size.

Figure 7.8 demonstrates path planning for 200 agents in a randomly generated environment of size 512 × 512. Since our approach can efficiently handle dynamic updates, we can interleave planning with execution to create a crowd simulator.

7.2   GPU-BASED DYNAMIC SEARCH ON ADAPTIVE RESOLUTION GRIDS

Section 7.1 describes a wavefront-based planner that takes advantage of the GPU to achieve great performance improvements. However, the approach is limited to uniform grid representa-

Table 7.2: Algorithm performance for different environments (time in seconds)

World Size	GPU 1		GPU 2	
	Exit A	Exit B	Exit A	Exit B
32 × 32	0.011	0.012	0.01	0.012
64 × 64	0.024	0.027	0.017	0.022
128 × 128	0.107	0.146	0.078	0.096
256 × 256	0.608	0.871	0.349	0.542
512 × 512	4.219	6.24	2.691	3.816
1024 × 1024	32.931	49.126	21.246	30.778
2048 × 2048	258.88	387.35	178.794	264.373

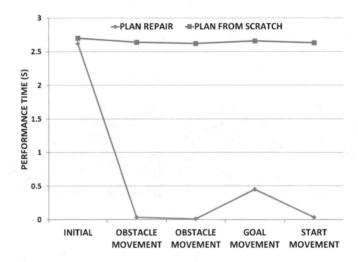

Figure 7.6: GPU planner performance for dynamic simulation with changes in environment, start, and goal.

tions of the environment, and a separate map needs to be maintained for each target location, resulting in a substantial memory and computational overhead. This makes the current approach unsuitable for agents with their own targets. Adaptive representations of the environment such as quad-trees reduce the memory overhead but require expensive queries which are not amenable for GPU operations. We address these limitations by representing the state space as quad-trees and performing adjacency detection using quad-codes.

By adaptively discretizing the environment into variable resolution states, finer resolution is used only where needed, thus significantly reducing the size of the state space. This approach sub-

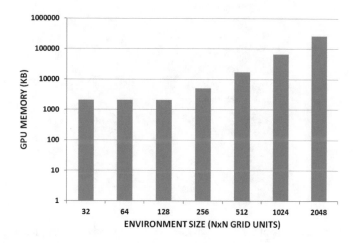

Figure 7.7: Memory usage in GPU for different environment sizes.

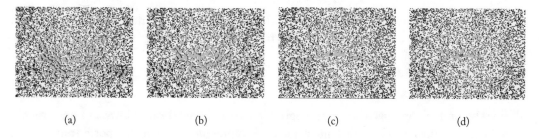

Figure 7.8: Global path planning and simulation of 200 agents on a complex navigation benchmark.

stantially decreases the memory requirements to represent an environment as well as accelerates the search process, but the resulting planner is no longer able to guarantee optimality.

Using an adaptive environment representation on the GPU has two main challenges: indexing is no longer constant time and handling dynamic changes is computationally expensive, since the number and location of neighboring states vary as obstacles move in the state space. Given the dynamic nature of quad-trees, it is impossible to know ahead of time how many neighbors a given state has, which ones those neighbors are, and how many states are needed to represent the state space.

In addition, dynamically allocating memory on the GPU to accommodate these changes can be very expensive. We address these challenges by: (1) using a *quadcode* scheme to perform efficient indexing, update, and adjacency detection operations on the GPU; and (2) efficiently handling dynamic environment changes by performing local repair operations on the quad-tree, and performing plan repair to resolve state inconsistencies without having to plan from scratch.

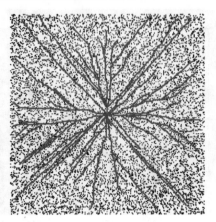

Figure 7.9: Illustration of our planning algorithm operating in large complex environments. The figure shows a grid world of 512×512 with 200 agents. The goal is in the center of the map, and the computed paths are shown in blue.

This section highlights the benefits of this new method on a variety of challenging benchmarks with empirical results and discusses the trade-offs between using an adaptive-resolution vs. a uniform grid.

7.2.1   METHOD OVERVIEW

The idea behind this approach is fairly straight forward and it works as follows: as a pre-processing step, the environment is subdivided into a quad-tree representation, which is ported onto the GPU to compute an initial plan. Environment changes triggers local repairs in the quad-tree and GPU memory is updated to reflect environment changes. The plan is finally repaired by updating only the costs which have been invalidated as a result of these changes.

A hierarchical representation of the environment for GPU computations presents several challenges. Traversing quad-trees is an inherently recursive operation which makes traditional tree-recursion techniques unsuited for the GPU. To further complicate things, the number of neighbors for a given of a quad cannot be known in advance, presenting a new challenge to use a parallel wavefront planner as in the previous section. To address these challenges, we use an efficient *quad-code* method for performing indexing, update, and neighbor finding operations. Quad-codes can be efficiently and independently computed using simple arithmetic operations, which makes them amenable for GPU processing. The main components of our system are enumerated below, and elaborated in subsequent sections: Figure 7.10 provides an overview of the GPU-based planning framework using adaptive-resolution grids on a simple benchmark.

- **Initialization.** Data structures used to represent the environment, GPU memory allocation, and indexing to reduce the performance impact of dynamically changing environments.

- **Plan Computation.** Wavefront propagation on an adaptive resolution environment representation, with minimal number of GPU iterations for plan computation.

- **Plan Repair.** Efficient plan repair to accommodate dynamic changes in the environment.

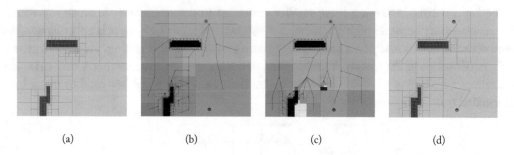

(a)   (b)   (c)   (d)

Figure 7.10: Method overview. (a) Initial subdivision of the environment into quads. (b) Plan computed for the top goal (red circle) for any quad. The arrows point to the predecessor quad and the colors correspond to each quad's g-value relative to a scale where $green = 0$, $blue = \max(g)$ and $white =$ not computed. (c) Quads locally repaired after obstacle movement from the previous image. (d) Plan computed for all agents.

7.2.2   METHOD INITIALIZATION

The environment is initially subdivided using a quad-tree up to a predefined maximum depth, d_{max}, which prevents the tree from growing to the point of practically becoming a uniform grid in highly populated environments. All leaf nodes (which correspond to quad regions) in the environment $\mathbb{Q}$, the number of quad divisions n_q, and the number of goals n_g are transferred to the GPU by calling **generateMap**$(\mathbb{Q}, n_q, n_g)$ and **createHashMap**$(\mathbb{Q}, n_q)$.

The function **generateMap** allocates memory for n_q quad structures and $n_q \times n_g$ cost structures in host memory. A quad structure is defined as $Q = \langle (x,y), cost, code, index \rangle$, where (x, y) is the center position, $cost$ is the cost of transitioning to this quad, $code$ is a unique quad identifier, and $index$ is the position the at which the quad is located in GPU. The $code$ is used as the key into a hashmap to retrieve the quad structure, while $index$ indexes into the computed costs for all goals with respect to this quad.

As presented before for the uniform grid, a different copy of the state space is created for each goal. Each copy is represented as an array of costs, and the cost associated with a particular goal for the same quad can be retrieved by using $index$ in the corresponding cost array. A cost structure is defined as $\langle g, pred\_code \rangle$, where $pred\_code$ is the code of the neighbor quad q' with minimum g-value, and g is computed as:

$$g(q) = g(q') + c(q,q') \cdot cost(q),$$

where $cost(q, q')$ is the Euclidean distance between the centers of q and q', and $cost(q)$ is the cost of transitioning to q.

The reason for creating two structures is twofold. First, creating several smaller representation of a given state saves memory when dealing with multiple goals (many attributes are shared for a given quad regardless of the goal), since only one copy of the whole quad structure needs to be kept. Second, accessing these smaller structures in device memory is much more efficient because memory becomes highly coalesced. As shown in Figure 7.11, there is only one quad structure per state and as many cost structures as there are goals. In this setting, propagating the wave only needs access to the cost structures instead of the entire quad. The method **createHashMap** initializes a hash map in device memory where the quad *code* is the key used to index into the associated quad structure.

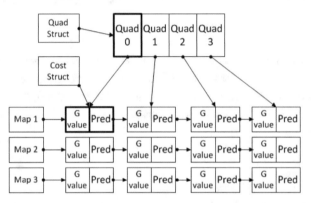

Figure 7.11: Quad struct and cost struct layout. For each quad, there are as many cost structs as goals in the map.

The final initialization step is to compute the neighbors for each quad. Given a predetermined maximum number of neighbors per quad, n_{max}, (we used $n_{max} = 20$), we allocate an array large enough to be able to support our needs throughout the lifetime of the planner. This array contains the indexes at which each neighbor is located in the cost struct array. An index of -1 means that no neighbor has been computed at that position.

The neighbors for each quad are computed as described in [Li and Loew, 1987a], and the planner is run to generate the initial plan. Figure 7.12 shows how quads maintain their neighbors' locations. We assume a simple subdivision where the spatial region is divided into seven quads, as shown in the figure. Each node in the tree can calculate an offset, given its position, in the neighbors array and find all adjacent quads. Having a predefined maximum number of neighbors allows us to easily identify the starting offset that corresponds to each quad's neighbors. When propagating costs, all neighbors costs for a given quad can be found by accessing the costs corresponding to the neighbor indexes until an index of -1 is found.

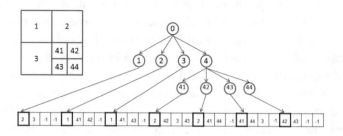

Figure 7.12: Mapping neighbor arrays to quads. The index of each quad corresponds to an offset on the neighbor array. A neighbor quad can be found using the quad code in the GPU hashmap.

7.2.3 ENVIRONMENT REPRESENTATION

The environment is subdivided by using a quad-tree with maximum depth d_{max}, in our tests $d_{max} = 7$ provided a good trade-off between plan quality and performance. During tree construction, a code is assigned to each quad in the following manner: the root of the tree is assigned 0 and it is always subdivided into four children. The top left child is assigned a code of 1, top right is assigned 2, bottom left child 3, and the bottom right child is assigned a code of 4. Each child node containing an obstacle is further subdivided into four child quads. Their code is equal to the code based on the relative location to their parent, appended to the parent's code.

Quad-codes are an efficient method for computing unique identifiers using simple arithmetic operations. They can be used for indexing, checking for quad existence, and neighbor finding in parallel. For more details on how quad-codes are computed, refer to [Li and Loew, 1987a,b].

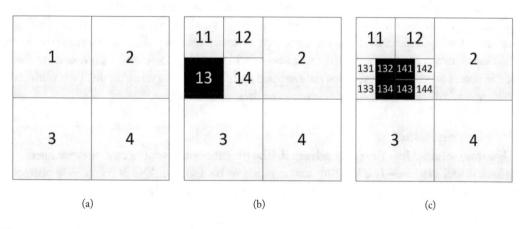

Figure 7.13: Quad-code computation for dynamic quads. The black obstacle subdivides the tree differently depending on its position. The numbers in each quad correspond to its code.

Figure 7.13 depicts a quad-tree representation for a simple state space containing a moving obstacle. In Figure 7.13(a) shows the initial tree subdivision where no obstacles are present. The root of the tree in four high-level quads with their corresponding quad-codes. In 7.13(b) an obstacle enters the region defined by quad 1, thus dividing it into four quads with their corresponding codes. This results in quads 11, 12, 13, and 14. The obstacle moves again in 7.13(c) intersecting quads 13 and 14, requiring each of them to subdivide into four more quads.

At the end of the process, a reference for each quad is inserted into the hashmap residing in GPU memory. The quad code serves as the key and the index in the host memory array is the value. The quads themselves are stored in CPU memory. For large environments, a given quad can contain multiple obstacles for it cannot be divided further than d_{max}. In these scenarios, rather than considering a quad as either an obstacle or a free space, we compute its cost as the relative area that is occupied by obstacles, producing high g-values for highly populated quads. This allows the planner to favor a path through empty quads. If a quad area is occupied by obstacles more than a given threshold (we used 90%), the entire quad is deemed untraversable.

7.2.4 DYNAMIC SEARCH ON THE GPU FOR ADAPTIVE RESOLUTION GRIDS

The same underlying approach used with uniform grids is applied here, where individual states represent quads of varying sizes instead of a grid cell. The propagation of the wave accounts for the cost of transitioning the quad as well as its area. At the end of the propagation process, the minimum cost of reaching any quad center from the goal is computed which is used to efficiently trace paths to the goal from any position in the environment.

Main Kernel

The main kernel creates two copies of the entire map of cost structs in device memory, one is used as read-only and the other one as write-only to avoid synchronization issues when running the kernel on multiple threads. It also creates a copy in device memory of all quad structures to access the data that is common to all cells regardless of which goal is associated with them. At the end of the kernel execution, these maps are swapped and the kernel is run again until the termination condition is satisfied and the costs converge to their optimal values.

Wavefront Expansion

Algorithm 21 describes the **plannerKernel**. The planner initializes the cost of every quad q to a default value, $g(q) = -1$, indicating that it needs to be updated. Quads containing obstacles o take a transition value relative to the total area occupied by obstacles:

$$cost(q) = \texttt{MAX\_COST} \cdot \frac{\sum_{\forall o \in q} \textbf{area}(o)}{\textbf{area}(q)},$$

Algorithm 20 computePlan(S_{cpu})

$S_r \leftarrow S_{cpu}$;
$S_w \leftarrow S_{cpu}$;
repeat
   $f \leftarrow 0$;
   plannerKernel($quads, neighbors, S_r, S_w, f$);
   swap (S_r, S_w);
until ($f = 0$);
$S_{cpu} \leftarrow S_r$;

---

where `MAX_COST` indicates that the entire quad is occupied by an obstacle. The goal quad is initialized with a value of 0, $g(\text{goal}) = 0$. The planner finds the value g of reaching any quad q from the *goal* by launching a kernel at each iteration that computes $g(\text{q})$ as follows:

$$g(q) = \min_{q' \in succ(q) \wedge g(q') \geq 0} (g(q') + c(q, q') \cdot cost(q)),$$

for each successor $q' \in succ(q)$ that has been updated (i.e., $g \geq 0$), where $cost(q, q')$ is the Euclidean distance between the centers of q and q', and $cost(q)$ is the cost of traversing q. This process continues until the minimum g-value expanded in an iteration is larger than the g-value found for the agent quad.

We utilize two copies of the map representation (arrays of cost structs): a write-only state space S_w and a read-only state space S_r, similar to the approach taken in the previous section. We also pass to the kernel the array of quad structs to access quad information that is goal independent, and an array containing all the precomputed neighbor indexes. Each thread, t_q, in the kernel reads corresponds to a given quad q.

Minimal Map Convergence. As discussed in the previous section, it is not necessary to run the planner until the entire map costs converge to their optimal values. We can use the exit conditions described before and achieve similar behavior in this new state space representation. Lines 9 and 10 of Algorithm 21 show the exit condition for early termination. The error of the plan found by using an adaptive resolution environment scheme depends on the size of the quads found as part of the solution.

Plan Repair
Efficiently repairing the quad-tree and updating the GPU hashmap is imperative to handle dynamic environments. A quad-tree can be locally repaired as follows: upon an obstacle movement, the parent of the quad it previously occupied is observed. If the parent does not contains an obstacle, all of its child nodes are removed and recursively repeat the process for its parent. Then, the sub-trees corresponding to each of the quads the obstacle now occupies are constructed.

Algorithm 21 plannerKernel$(quads, neighbors, m_r, m_w, flag)$

1:   $q \leftarrow quad(t_q)$;
2: **if** $(q \neq \texttt{INVALID} \wedge q \neq goal)$ **then**
3:     **for each** $q' \in neighbor(q)$ **do**
4:       **if** $(index(q') \geq 0)$ **then**
5:         $g_{new} \leftarrow g(q') + c(q, q') \cdot cost(q')$;
6:         **if** $((g_{new} < g(q) \vee g(q) = -1) \wedge g(q') > -1)$ **then**
7:           $pred\_code(q) \leftarrow code(q')$;
8:           $g(q) \leftarrow g_{new}$;
9:           **if** $(g(s) < g(start) \vee g(agent) = -1)$ **then**
10:             $flag = 1$;
11:           **end if**
12:         **end if**
13:       **end if**
14:     **end for**
15: **end if**

As the tree is repaired, we keep track of which quads have been added, removed, and the ones that need to be updated. A quad may need to be updated when an obstacle movement invalidates its current g-value, requiring its cost to be recomputed. The list of quads that have been inserted, removed, and need to be updated, are processed on the GPU as follows.

1. For each quad to be updated, we compute its quadcode to find its index by querying the hashmap and invalidate its cost values by setting them to -1, which mandates a recomputation the next time the planner kernel is launched.

2. A kernel is launched to account for all removed quads which retrieves their indices and invalidates them in the hash map, as well as its corresponding g-values in the cost map. Additionally, all references to invalidated quads in the precomputed neighbor list is detected and removed.

3. We insert the new quads into our quads array and cost map. We recycle memory that is occupied by invalidated quads that were previously removed, or append the newly created quads at the end of the list. We also keep track of the locations where these new quads were inserted and launch a kernel that updates the hashmap with the indices and quadcodes for each quad.

4. A kernel is launched (Algorithm 22) which computes the neighbors of each inserted quad, and updates the neighbors of surrounding quads, if needed.

Algorithm 22 UpdateNeighborsKernel($inserted$, $hashmap$)

$q \leftarrow threadQuad$;
$computeNeighbors(q)$;
for each $q' \in neighbors(q)$ **do**
  if $(\text{code}(q') > 0)$ **then**
    $index \leftarrow hashmap[\text{code}(q')]$;
    if $(index \geq 0 \wedge q' \notin inserted)$ **then**
      $computeNeighbors(q')$;
    end if
  end if
end for

An obstacle movement could leave a quad q in an inconsistent state. A quad is considered inconsistent when

$$g(q) \neq g(q') + c(q, q') \cdot cost(q).$$

Repairing Inconsistent States. After updating the neighbors, we fix any inconsistency in the state space and propagate this repair until there are no inconsistent states left, as described in Algorithm 23. The algorithm runs in a loop which sets the flag $propagateUpdate$ to 0 before running the kernel. The loop is repeatedly executed until no changes are made in the most recent kernel execution (i.e., the flag is not modified). The kernel checks if the g-value of a quad is inconsistent or its predecessor had been invalidated. If so, the quad g-value is set to -1 and its $pred\_code$ is set to 0, which marks it for update the next time the planner kernel is launched,

---

Algorithm 23 UpdateAfterObstacleMove($quads$, $hashmap$, $propagateUpdate$)

$q \leftarrow threadQuad$;
if $(\text{pred\_code}(q) > 0)$ **then**
  $predIndex \leftarrow hashmap[\text{pred\_code}(q)]$;
  if $(predIndex < 0 \vee g(q) \neq g(q') + c(q, q') \times cost(q))$ **then**
    $g(q) = -1$;
    $\text{pred\_code}(g) = 0$;
    $propagateUpdate = 1$;
  end if
end if

7.2.5   MULTI-AGENT AND MULTI-GOAL PLANNING

The adaptive resolution planner discussed in this section is able to handle multiple agents while scaling to handle larger environments and multiple goals. The results that follow use RVO2 [Berg et al., 2011] for local collision avoidance and interleave planning and execution on background threads.

A copy of the environment costs map is created for each goal. The adaptive representation of the state space greatly reduces the running time to find a plan for all agents. By simply assigning an agent to a different cost map from the one it was originally associated with, each agent is able to change its target to any other goal with minimal effort. This is particularly useful for simulating large crowds of agents that travel between predefined locations in the environment.

Path Smoothing

The implementation of the system shown searches for neighbors in the north, west, east, and south directions. The plan quality can be improved by considering diagonal directions as well. To improve path smoothness, we perform a raycast from the agents current position to its next set of waypoints along the path to check for the farthest waypoint that does not have an obstacle along a straight-line path. This allows us to disregard waypoints along the path which would otherwise produce jagged agent movements. This is an example of one kind of post-processing smoothing technique that can be used, and there are also other smoothing techniques which can be employed to further improve the quality of the simulations.

7.2.6   RESULTS

This section demonstrates the benefits of an adaptive state representation by testing several challenging navigation benchmarks. The experiments were performed on different environments sizes: 128×128, 256×256, 512×512, $1024 \times 1,024$, and $2,048 \times 2,048$. All sizes are given in units, where a unit maps to a single grid cell in a uniform grid. Figure 7.14 shows the number of states required to represent the state space using a quad-tree of depth 7 vs. a uniform grid. On a 128×128 world map, the uniform grid needs $16,384$ states to represent the map while a quad-tree representation only requires $1,303$ states. This difference keeps increasing with larger sizes, the greatest difference being for a map of size $2,048 \times 2,048$, where a uniform grid requires $4,194,304$ states while only $21,312$ is required for a quad-tree. It is worth noting that the number of states for a uniform grid depends directly on the size of the environment, while the number of quads in the quad-tree also depends on how obstacles are distributed throughout the environment.

Figure 7.15 compares GPU memory usage for different world sizes using different tree depth. For the adaptive representation approach, the planner requires some extra memory compared to the uniform grid. This is due to the pre-computed list of neighbors and hashmap that needs to be stored in GPU memory, which is not required for the uniform grid. Starting at a grid size of 512×512 we observe the benefits of representing the state space with quad-trees. For quad-trees with a maximum depth of 5 and 7, the lack of resolution yields a final plan that is not as accurate. For a quad-tree with depth 9, the resulting plan resembles the result of the uniform grid while utilizing significantly fewer resources.

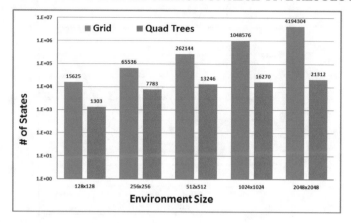

Figure 7.14: Comparison of number of states required to represent the same map using a uniform and adaptive grid. For a world of 128×128, an adaptive grid need about 10 times fewer states, while for a world of 2048×2048 it utilizes about 200 times fewer states.

The graph shown in figure 7.16 compares the running time between quad-trees and uniform grid for computing an initial plan and performing several repairs due to dynamic changes in the environment. The adaptive environment representation outperforms the uniform grid with significant performance gains achieved during initial plan computation.

It could be the case that an obstacle movement does not affect a plan in a uniform grid, but it does when using quad-trees. If a change forces a quad that was part of the solution to subdivide, this will require some effort to repair the plan, while the same change may not affect a uniform grid. This situation causes the uniform grid to outperform the quad-tree in the last repair.

Table 7.3: Algorithm performance for different environments (time in seconds)

World Size	GPU		
	Uniform Grid	Quad Tree - depth 7	Reduction as %
128×128	0.9	0.012	1.3
256×256	1.25	0.04	3.2
512×512	10.8	0.187	1.73
1024×1024	14.12	0.04	0.28
2048×2048	196.14	0.234	0.11

Table 7.3 compares the running time of the different test scenarios for the uniform and adaptive grid. The last column shows the running time of the quad-tree as a percentage of the uniform grid. From the data we can observe that the world size is not the major determinant of the required running time for quad trees, instead it mostly depends on obstacle distribution. We

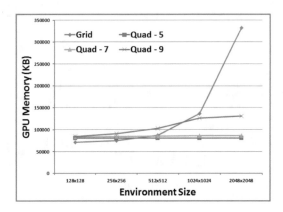

Figure 7.15: GPU memory used for different world sizes. We compare results between a uniform grid and a quad-tree of depths 5, 7, and 9.

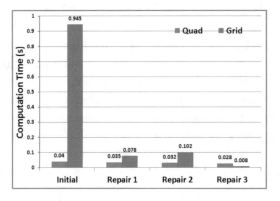

Figure 7.16: Comparison of running times between uniform grid and quad-trees of the initial plan and several repairs in a map of size 512×512 units.

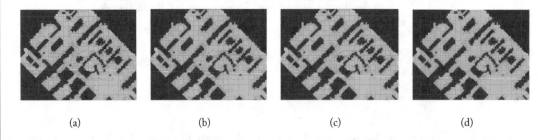

(a)   (b)   (c)   (d)

Figure 7.17: Several frames of a multi-agent simulation in a game environment. The environment is not axis-aligned which is handled by the quad subdivision of the environment.

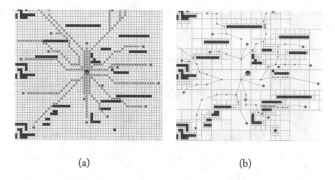

(a)   (b)

Figure 7.18: Comparison of paths obtained using an uniform grid (a) vs. an adaptive resolution grid (b). Red dots mark each waypoint in the final plan. (a) Retrieves optimal paths by searching many more states. (b) Sacrifices plan quality but results in enormous memory and performance gains.

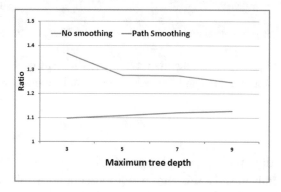

Figure 7.19: Comparison of path lengths in adaptive resolution with and without path smoothing vs. uniform grid. The figure shows the ratio of the average path lengths for all agents in various test cases with the average path lengths for the uniform grid.

observe a significant performance boost when using an adaptive grid, with a 1000X speedup for a world size of 2048 × 2048. In the worst case, the adaptive resolution approach will perform as poorly as a uniform grid, but this scenario rarely happens in practice. Figure 7.18 compares the solution obtained using a uniform and adaptive grid. Although plan quality is sacrificed by using a coarser environment representation, the performance benefits far outweigh the loss of quality. Figure 7.19 compares the path lengths of the adaptive grid in various scenarios with the uniform grid. We can observe that solutions given using quads is a little over 20% longer than the optimal solutions on average, but just above 10% when we use path smoothing.

Figure 7.20 shows the planner running with 250 agents distributed among 8 goals. The world size for this scenario is 1024 × 1024. On the machine used for testing, the uniform grid was not able to handle such a large problem while the quad planner returned a solution for all agents in under 4 s.

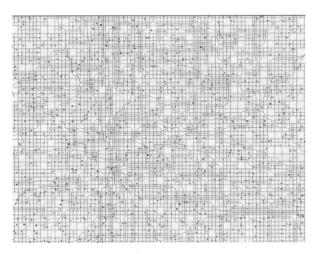

Figure 7.20: Computed plans for 250 agents traveling to different goals in a 1024×1024 environment. Lines show the computed plan and line color corresponds to different destinations.

7.3 SUMMARY

We explored the question of how to exploit the GPU to improve the running time to find a solution in planning problems. To this end, we proposed a massively parallel planner based on wave-front techniques, which efficiently handles world changes and agent movements by reusing previous search efforts. These properties makes it amenable for global path planning involving hundreds or thousands of agents in complex environments.

To address the prohibitive memory requirements especially for large search graphs, we used an adaptive environment representation. We have used efficient queries for quad indexing and neighbor finding via quad-codes, which facilitates parallelization. With these solutions in place, we are able to scale to handle large, complex, dynamic environments with thousands of agents with different targets. In addition to addressing these memory limitations, this planner provides a significant performance boost and still preserves dynamic properties.

An avenue of potential future exploration is to investigate the integration of hierarchical planning methods with anytime dynamic planners to further harness the potential of abstract environment representations. Porting these algorithms to the GPU is a challenging research direction that merits further investigation. Based on these results, we advocate that the GPU can provide great benefits in planning problems and it is well worth the effort designing a parallel approach to meet one's needs. When leveraging all computations to the GPU is not appropriate, a hybrid approach may still yield substantial benefits and is worth investigating depending on the needs of the application. Planning for groups of agents simultaneously [Huang et al., 2014] increases the dimensionality of the planning problem and may also benefit from massive parallelization.

PART IV

Planning Techniques for Character Animation

CHAPTER 8

Dynamic Planning of Footstep Trajectories for Crowd Simulation

The important ability of autonomous virtual characters to locomote from one place to another in dynamic environments has generally been addressed as the combination of two separate problems: (a) navigation and (b) motion synthesis. For the most part, research along these two directions has progressed independently of one another. The vast majority of navigation algorithms use a trivial locomotion model, outputting a simple force or velocity vector and assuming that motion synthesis can realistically animate a character to follow these arbitrary commands. This simplistic interface between navigation and motion synthesis introduces important limitations.

- *Limited locomotion constraints:* Very few navigation algorithms account for locomotion constraints. Trajectories may have discontinuous velocities, oscillations, awkward orientations, or may try to move a character during the wrong animation state, and these side-effects make it harder to animate the character intelligently. For example, a character cannot easily shift momentum to the right when stepping with the right foot, and a character would rarely side-step for more than two steps at a steady walking speed.

- *Limited navigation control:* Existing navigation systems usually assume that motion synthesis will automatically know when to perform subtle maneuvers, such as side-stepping vs.

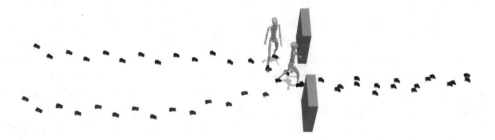

Figure 8.1: Autonomous characters navigating with our footstep locomotion model. Footstep navigation allows characters to maneuver with spatial precision, temporal precision, and natural locomotion.

reorienting the torso, stepping backward vs. turning around, stopping and starting, planting a foot to change momentum quickly, or carefully placing footsteps in exact locations. However, there are many cases where navigation should be aware of such options and have some control over their execution. For example, a character standing next to a wall would probably turn around toward the open space, not toward the wall. A character could better navigate through tight spaces, including dynamic spaces between other navigating characters, if its collision radius were adaptable.

Existing systems rely on robust motion synthesis to address such problems, but this forces motion synthesis to make undesirable trade-offs: either the motions follow navigation commands precisely, which can cause discontinuous or awkward animations, or the motions ensure proper, natural locomotion regardless of the requested timing, which can cause delayed reactions and collisions. In either case, the navigation intelligence of the character is diminished in the final result that a viewer sees. For example, many high quality motion synthesis techniques based on motion graphs have a latency between receiving and executing a command, because of the time it takes to continue animating until a valid transition is possible. These issues motivate the need for navigation to be more locomotion-aware and the need for better integration between navigation and motion synthesis.

We present one possible solution to this problem: generating sequences of footsteps as the interface between navigation and motion synthesis. Foot placement and timing is an intuitive and representative abstraction for most locomotion tasks. Footsteps offer finer control which allows navigation to choose more natural and nuanced actions. At the same time, the series of footsteps produced by navigation communicates precise, unambiguous spatial and timing information to motion synthesis. We use a space-time planning approach to dynamically generate a short-horizon sequence of footsteps.

Footsteps are represented as a 2D locomotion trajectory that approximates an inverted spherical pendulum model. The pivot of the pendulum trajectory represents a foot's location and orientation. The character navigates by computing an efficient sequence of space-time trajectories (footsteps) that avoids time-varying collisions, satisfies footstep constraints for natural locomotion, and minimizes the effort to reach a goal. In its most general form, this is a difficult nonholonomic optimization planning problem in continuous space where the configuration and action spaces must be dynamically re-computed for every plan [LaValle, 2006]. We mitigate these challenges by: (a) using an *analytic* approximation of inverted pendulum dynamics, (b) allowing near-optimal plans, (c) exploiting domain-specific knowledge, and (d) searching only a limited horizon of footsteps. Our approach generates natural footsteps in real-time for hundreds of characters. Characters successfully avoid collisions with each other and produce dynamically stable footsteps that correspond to natural and fluid motion, with precise timing constraints. Because the most significant biomechanics and timing constraints are already taken into account in the navigation locomotion model, using a motion synthesis algorithm that follows footsteps

is straightforward and results in animations that are nuanced, collision-free, and plausible. For previous publications on this topic, please refer to [Singh et al., 2011a,b].

8.1   LOCOMOTION MODEL

The primary data structure in our locomotion model is a *footstep*, storing: (1) the trajectory of the character's center of mass, including position, velocity, and timing information, (2) the location and orientation of the foot itself, and (3) the cost of taking the step. We next describe how the model computes these aspects of a footstep, as well as the constraints associated with creating a footstep.

8.1.1   INVERTED PENDULUM MODEL

Our approach is inspired by the *linear inverted pendulum* model [Kajita et al., 2001] which produces equations of motion derived by constraining the center of mass to a horizontal 2D plane. The passive dynamics of a character's center of mass pivoting over a footstep located at the origin is then described by:

$$\ddot{x}(t) = \frac{g}{l}x, \tag{8.1}$$

$$\ddot{y}(t) = \frac{g}{l}y, \tag{8.2}$$

where g is the positive gravity constant, l is the length of the character's leg from the ground to the center of mass (i.e., the length of the pendulum shaft), and (x, y) is the time-dependent

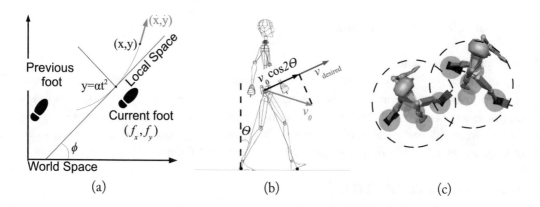

Figure 8.2: Our footstep locomotion model. (a) Depiction of state and footstep action parameters. The state includes the 2D approximation of an inverted spherical pendulum trajectory. (b) A sagittal view of the pendulum model used to estimate energy costs. (c) The collision model uses five circles that track the torso and feet over time, allowing tighter configurations than a single coarse radius.

2D position of the character's center of mass (i.e., the pendulum weight). Note that both x and y have the same dynamics as the basic planar inverted pendulum when using the small-angle approximation, $x \simeq \sin x$. This is valid for the region of interest since the angle between a human leg and the vertical axis rarely exceeds 30°.

Here we further assume that l remains constant—a reasonable approximation within the small-angle region. Then, given initial position (x_0, y_0) and initial velocity (v_{x_0}, v_{y_0}), the solution for $x(t)$ is a general hyperbola:

$$x(t) = \frac{(kx_0 + v_{x_0})e^{kt} + (kx_0 - v_{x_0})e^{-kt}}{2k}, \tag{8.3}$$

and $y(t)$ has a similar solution, where $k = \sqrt{\frac{g}{l}}$.

We use the second-order Taylor series approximation of this parabola:

$$x(t) = x_0 + v_{x_0}t + \frac{x_0 k^2}{2}t^2, \tag{8.4}$$

$$y(t) = y_0 + v_{y_0}t + \frac{y_0 k^2}{2}t^2. \tag{8.5}$$

Finally, this parabola is translated, rotated, reflected, and re-parameterized from world space into a canonical parabola in local space:

$$x(t) = v_{x_0}t, \tag{8.6}$$
$$y(t) = \alpha t^2, \tag{8.7}$$
$$\dot{x}(t) = v_{x_0}, \tag{8.8}$$
$$\dot{y}(t) = 2\alpha t, \tag{8.9}$$

such that both v_{x_0} and α are positive.

Evaluating the trajectory

Equations (8.6)–(8.9) allow us to *analytically* evaluate the position and velocity of a character's center of mass at any time. The implementation is extremely fast: first, the local parabola and its derivative are evaluated at the desired time, and then the local position and velocity are transformed from the canonical parabola space into world space. This makes it practical to search through many possible trajectories for many characters in real-time.

8.1.2   FOOTSTEP ACTIONS

The state of the character $s \in \mathbf{S}$ is defined as follows:

$$s = \langle \vec{x}, (\dot{x}, \dot{y}), (f_x, f_y), f_\phi, I \in \{\texttt{left}, \texttt{right}\} \rangle, \tag{8.10}$$

where $\vec{x}$ and $(\dot{x}, \dot{y})$ are the position and velocity of the center of mass of the character at the *end* of the step, (f_x, f_y) and f_ϕ are the location and orientation of the foot, and I is an indicator of which foot (left or right) is taking the step. The state space $\mathbf{S}$ is the set of valid states that satisfy the constraints of the locomotion model described below.

A *footstep action* transitions a character from one state to another. An action a $\in \mathbf{A}$ is given by:

$$a = \langle \phi, v_{\text{desired}}, T \rangle, \qquad (8.11)$$

where ϕ is the desired orientation of the parabola, v_{desired} is the desired initial speed of the center of mass, and T is the desired time duration of the step. The action space $\mathbf{A}$ is the set of valid footstep actions, where the input and output states are both valid.

A key aspect of the model is the transition function, $s_n = \textbf{createFootstep}(s, a)$. This function receives a desired footstep action a and a state s and returns a new state s_n if the action is valid. It is implemented as follows. First, ϕ, which indicates the orientation of the parabola, is used to compute a transform from world space to local parabola space. Then, the direction of velocity $(\dot{x}, \dot{y})$ from the end of the previous step is transformed into local space, normalized, and rescaled by the desired speed v_{desired}. With this local desired velocity, there is enough information to solve for α, and then Equations (8.6)–(8.9) are used to compute $\vec{x}$ and $(\dot{x}, \dot{y})$ at the end of the next step. The foot location is always located at $(f_x, f_y) = (0, -d)$ in local space, where d is a user-defined parameter that describes the distance between a character's foot and center at rest. Additionally, for every footstep, an interval of valid foot orientations is maintained, and the exact foot orientation within this interval is chosen with a fast, simple post-process. (We will discuss foot orientations below.) Finally, all state information is transformed back into world space, which serves as the input to create the next footstep. During the entire process, checks are made to verify that the transition and new state are valid based on constraints, described next.

8.2   LOCOMOTION CONSTRAINTS

We introduce several constraints on footstep selection to ensure that the steps chosen by the agents satisfy orientation constraints, and avoid collisions with other characters.

8.2.1   FOOTSTEP ORIENTATION

Intuitively, it would seem that footstep orientations must be an additional control parameter when creating a footstep. However, the choice of footstep orientation seems to have no direct effect on the dynamics of the center of mass trajectory; in our experience the foot orientation only constrains the options for future trajectories and future footsteps. To implement this constraint, we compute an interval $[f_{\phi\text{inner}}, f_{\phi\text{outer}}]_{next}$ of valid foot orientations when creating a footstep. This interval is constrained by the same interval from the previous step, and further constrained by the choice of parabola orientation ϕ used to create the next footstep:

$$[f_{\phi\text{inner}}, f_{\phi\text{outer}}]_{next} = [f_{\phi\text{outer}}, f_{\phi\text{inner}} + \frac{\pi}{2}] \cup [\phi, (\dot{x}, \dot{y})]. \qquad (8.12)$$

In other words, the constraining interval $[\phi, (\dot{x}, \dot{y})]$ describes that the character would not choose a foot orientation that puts its center of mass on the outer side of the foot. The ordering of bounds in these intervals insures that the next foot's outer bound is constrained by the previous foot's inner bound. Similarly, a human would rarely orient its next footstep more outwards than the orientation of their momentum—this constraint is unintuitive because it is easy to overlook momentum. Finally, parameters can be defined that allow a user to adjust these constraints, for example, to model a slow character that has difficulty with sharp, quick turns.

8.2.2   SPACE-TIME COLLISION MODEL

For any given footstep, this model can compute the *time-varying* collision bounds of the character at any exact time during the step. Thus, to determine if a footstep may cause a collision, we iterate over several time-steps within the footstep and query the collision bounds of other nearby characters for that time. The collision bounds are estimated using 5 circles, as depicted in Figure 8.2(c). If a potential footstep action causes any collision, that footstep is considered invalid. For additional realism, a character can only test collisions with other characters that overlap its visual field.

8.2.3   USER PARAMETERS

Our locomotion model offers a number of intuitive parameters that a user can modify. These parameters include the preferred value and limits of the duration of the step, the preferred value and limits on the stride length, the desired velocity of the center of mass and its limits, the interval of admissible foot orientations described above and its bounds, etc. By modifying these parameters a user can practically create new locomotion models. For example, restricting the range of some of the parameters results in a locomotion model that may reflect motion limits associated with the elderly.

Our model uses common default values for these parameters. For example, the desired velocity is set based on the Froude ratio heuristic, Fr. For a typical human-like walking gait, an average value of the Froude ratio is 0.25 [Biknevicius and Reilly, 2006]. The associated desired velocity is $v = \sqrt{glFr}$, where g is the acceleration of gravity and l is the height of the pelvis.

8.3   COST FUNCTION

We define the cost of a given step as the energy spent to execute a footstep action. We model three forms of energy expenditure for a step: (1) ΔE_1, a fixed rate of energy the character spends per unit time, (2) ΔE_2, the work spent due to ground reaction forces to achieve the desired speed, and (3) ΔE_3, the work spent due to ground reaction forces accelerating the center of mass along the trajectory. The total cost of a a footstep action transitioning a character from s to s_n is given by:

$$c(s, s_n) = \Delta E_1 + \Delta E_2 + \Delta E_3. \tag{8.13}$$

8.3.1   FIXED ENERGY RATE

The user defines a fixed rate of energy spent per second, denoted as P. For each step, this energy rate is multiplied by the time duration of the step T to compute the cost:

$$\Delta E_1 = P \cdot T. \tag{8.14}$$

This cost is proportional to the amount of time it takes to reach the goal: thus minimizing this cost corresponds to the character trying to minimize the time it spends walking to its goal. We found that good values for P are roughly proportional to the character's mass.

8.3.2   GROUND REACTION FORCES

We model three aspects of ground reaction forces that are exerted on the character's center of mass, based on the biomechanics literature [Kuo, 2007]. The geometry and notation of the cost model is shown in Figure 8.2. First, at the beginning of a new step, some of the character's momentum dissipates into the ground. We estimate this dissipation with straightforward trigonometry, reducing the character's speed from v_0 to $v_0 \cos(2\theta)$. Second, in order to resume a desired speed, the character actively exerts additional work on its center of mass, computed as:

$$\Delta E_2 = \frac{m}{2} \left| (v_{\text{desired}})^2 - (v_0 \cos(2\theta))^2 \right|. \tag{8.15}$$

This cost measures the effort required to choose a certain speed. At every step, some energy is dissipated into the ground, and if a character wants to maintain a certain speed, it must actively add the same amount of energy back into the system. On the other hand, not all energy dissipates from the system after a step, so if the character wants to come to an immediate stop, the character also requires work to remove energy from the system. Minimizing this cost corresponds to finding footsteps that require less effort, and thus tend to look more natural. Furthermore, when walking with excessively large steps, $\cos(2\theta)$ becomes smaller, implying that more energy is lost per step.

There is much more complexity to real bipedal locomotion than this cost model; for example, the appropriate bending of knees and ankles, and the elasticity of human joints can significantly reduce the amount of energy lost per step, thus reducing the required work of a real human versus this inverted pendulum model. However, while the model is not an accurate measurement, it is quite adequate for *comparing* the effort of different steps.

ΔE_2 captures only the cost of changing a character's momentum at the beginning of each step, but momentum may also change during the trajectory. For relatively straight trajectories, this change in momentum is mostly due to the passive inverted pendulum dynamics that requires no active work. However, for trajectories of high curvature, a character spends additional energy to change its momentum. We model this cost as the work required to apply the force over the length of the step, weighted by constant w:

$$\Delta E_3 = w \cdot F \cdot \text{length} = w \cdot m\alpha \cdot \text{length}. \tag{8.16}$$

Note that α is the same coefficient in Equation (8.7): the acceleration of the trajectory. α increases if the curvature of the parabola is larger, and also if the speed of the character along the trajectory is larger. Therefore, minimizing this cost corresponds to preferring straight steps when possible, and preferring to go slower (and consequently, take smaller steps) when changing the direction of momentum significantly. The weight w can be adjusted to prioritize whether the character prefers to stop (it costs less energy to avoid turning) or walk around an obstacle (it costs more energy to stop). In general we found good values of weight w to be between 0.2 and 0.5. The interpretation is that 20–50% of the curvature is due to the character's effort.

8.4   PLANNING ALGORITHM

Our autonomous characters navigate by planning a sequence of footsteps using the locomotion model described above. In this section we detail the implementation and integration of the short-horizon planner that outputs footsteps for navigation.

8.4.1   GENERATING DISCRETIZED FOOTSTEP ACTIONS

The choices for a character's next step are generated by discretizing the action space **A** in all three dimensions and using the **createFootstep**(s, a) function to compute the new state and cost of each action. We have found that v_{desired} and T can be discretized extremely coarsely, as long as there are at least a few different speeds and timings. Most of the complexity of the action space lies in the choices for the parabola orientation, ϕ. The choices for ϕ are defined relative to the velocity $(\dot{x}, \dot{y})$ at the end of the previous footstep, and the discretization ranges from almost straight to almost U-turns. The first choice that humans would usually consider is to step directly toward the local goal. To address this, we create a special option for ϕ that would orient the character directly toward its goal. With this specialized goal-dependent option, we found it was possible to give fewer fixed options for ϕ, focusing on larger turns. Without this option, even with a large variety of choices for ϕ, the character appears to steer toward an offset of the actual goal and then takes an unnatural corrective step.

8.4.2   SHORT-HORIZON BEST-FIRST SEARCH

Our planner uses a short-horizon best-first search algorithm that returns a near-optimal path, $P_k^s = \{s_{\text{start}}, s_1, s_2 \ldots s_k\}$, from a start state s_{start} toward a goal state s_{goal}.

The cost of transitioning from one state to another is given by $c(s, s_n)$, described by Equations (8.13)–(8.16). The heuristic function, $h(s)$ estimates the energy along the minimal path from s to s_{goal}:

$$h(s) = c_{min} \times \min |P_g^c|, \tag{8.17}$$

where c_{min} is the minimum energy spent in taking one normal footstep action, and $\min |P_g^c|$ is the number of steps along the shortest distance to the goal.

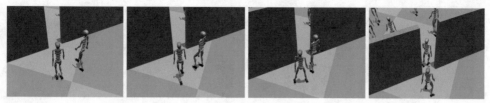

Figure 8.3: A character side-steps and yields to the other pedestrian, and then precisely navigates through the narrow doorway.

The path returned is complete, i.e., $s_k = s_{goal}$ if s_{goal} lies within the horizon of the planner. The horizon of the planner, N_{max} is the maximum number of nodes that can be expanded by the planner for a single search. As $N_{max} \to \infty$, the path returned by the planner is complete. For efficiency reasons, however, we limit the value of N_{max} to reasonable bounds. If the planner reaches the maximum limit of nodes to be expanded without reaching the goal, it instead constructs a path to a state from the closed list that had the best heuristic. Intuitively, this means that if no path is found within the search horizon, the planner returns a path to the node that has the most promise of reaching the goal.

8.5   EVALUATION

Table 8.1: Performance of our footstep planner for a character. The typical worst case plan generated up to 5000 nodes.

	Egress 200 agents	2-way traffic 200 agents	700 obstacles 500 agents
Avg. # nodes generated	137	234	261
Avg. # nodes expanded	82	190	192
Planner performance	1.6 ms	4.4 ms	3 ms
Amortized cost 20 Hz	0.037 ms	0.1 ms	0.11 ms

Performance and search statistics are shown in Table 8.1. One reason that our planner is fast is because of the scope of footsteps: a short horizon plan of 5–10 footsteps takes several seconds to execute but only a few milliseconds to compute. The amortized cost of updating a character at 20 Hz is shown in the table.

The space-time aspect of our planner helps the character to exhibit predictive, cooperative behaviors. It can solve challenging situations such as potential deadlocks in narrow spaces, e.g., Figure 8.3 depicts a character side stepping to allow the other pedestrian to pass first. Because the doorway is barely wide enough to fit a single pedestrian, many other navigation techniques would rely on collision prevention at the walls until the character eventually finds the doorway.

Figure 8.4: An egress simulation. Characters do not get stuck around the corners of the glass door.

Our time-varying collision model, as shown in Figure 8.2, dynamically adjusts the bounds of a character more realistically than fixed size disks, and thus allows much tighter spacing in crowded conditions. In Figure 8.4 a group of characters tightly squeeze through a narrow door. The planner and locomotion model offer a number of intuitive and fairly detailed parameters to interactively modify agent behaviors.

8.5.1 INTERFACING WITH MOTION SYNTHESIS

Interfacing with this footstep navigation method is very flexible. It can output properly timed footsteps, center of mass trajectories and velocities at any given time. Of course, to take advantage of the full intelligence of our approach, the motion synthesis should ideally be able to follow properly timed sequences of footsteps, including footstep orientation. For offline, high-quality animation systems, our method can be modified to output more accurate center of mass trajectories by using a hyperbola described in Equation (8.3) instead of the parabolic approximation, and by modifying the locomotion model and planner accordingly. The hyperbolic form is slower to evaluate, but it and its derivatives are still analytic, and the center of mass trajectories more accurately represent the inverted pendulum dynamics—in particular, the way a human's center of mass "lingers" at the apex of the motion. The work in [Beacco et al., 2015] describes a data-driven method for animating characters with foostep-level precision.

CHAPTER 9

Planning using Multiple Domains of Control

Once we can readily compute *where* an agent can step, the next task is to decide *how* to achieve its destination. Contemporary interactive applications require high fidelity navigation of interacting agents in non-deterministic, dynamic virtual worlds. The environment and agents are constantly affected by unpredictable forces (e.g., human input), making it impossible to accurately extrapolate the future world state to make optimal decisions. These complex domains require robust navigation algorithms that can handle partial and imperfect knowledge, while still making decisions which satisfy space-time constraints.

Different situations require different granularity of control. An open environment with no other agents and only static obstacles requires only coarse-grained control. Cluttered dynamic environments require fine-grained character control with carefully planned decisions with spatial *and* temporal precision. Some situations (e.g., potential deadlocks) may require explicit coordination between multiple agents.

The problem domain of interacting agents in dynamic environments is high-dimensional and continuous, with infinite ways to interact with objects and other agents. Having a rich action set, and a system that makes intelligent action choices, facilitates robust, realistic virtual characters, at the expense of interactivity and scalability. Greatly simplifying the problem domain yields interactive virtual worlds with hundreds and thousands of agents that exhibit only simple behaviors. The ultimate, far-reaching goal is still a considerable challenge: a real-time system for

Figure 9.1: Two agents navigating with space-time precision through a complex dynamic environment.

autonomous character control that can handle many characters, without compromising control fidelity.

We present a framework that uses multiple heterogeneous problem domains of differing complexities for navigation in large, complex, dynamic virtual environments. These problem domains (spaces of decision-making) differ in the complexity of their state representations and the fidelity of agent control. Domains range from a static navigation mesh which only accounts for static objects in the environment, to a space-time domain that factors in dynamic obstacles and other agents at much finer resolution. These domains provide different trade-offs in performance and fidelity of control, requiring a framework that efficiently works in multiple domains by using plans in one domain to focus and accelerate searches in more complex domains.

A global planning problem (start and goal configuration) is dynamically decomposed into a set of smaller problem instances across different domains, where an anytime dynamic planner is used to efficiently compute and repair plans for each of these problems. Planning tasks are connected by either using the computed path from one domain to define a *tunnel* to focus searches, or using successive waypoints along the path as start and goal for a planning task in another domain to reduce the search depth, thereby accelerating searches in more complex domains. Using this framework, we demonstrate real-time character navigation for multiple agents in large-scale, complex, dynamic environments, with precise control, and little computational overhead. This work is complementary to concurrent work [Gochev et al., 2011] that proposes the use of *tunnels* to accelerate searches in high-dimensional continuous domains. In contrast, this chapter leverages the explicit definition of multiple domains of control of varying levels of abstraction and proposes algorithms that can efficiently work across multiple domains simultaneously. This chapter is a summary of the findings previously described in [Kapadia et al., 2013].

9.1   OVERVIEW

The problem domain of a planner determines its effectiveness in solving a particular problem instance. A complex domain that accounts for many factors such as dynamic environments and other agents, and has a large branching factor in its action space can solve more difficult problems, but at a larger cost. A simpler domain benefits computational efficiency while compromising on control fidelity. Our framework enables the use of multiple heterogeneous domains of control, providing a balance between control fidelity and computational efficiency, without compromising either.

A global problem instance $\mathbf{P}_0$ is dynamically decomposed into a set of smaller problem instances $\{\mathbf{P}'\}$ across different planning domains $\{\Sigma_i\}$. Section 9.2 describes the different domains, and Section 9.4 describes the problem decomposition across domains. Each problem instance $\mathbf{P}'$ is assigned a planning task $T(\mathbf{P}')$, and an anytime dynamic planner is used to efficiently compute and repair plans for each of these tasks, while using plans in one domain to focus and accelerate searches in more complex domains. Plan efforts across domains are reused in two ways. The computed path from one domain can be used to define a *tunnel* which focuses the search, reducing its effective branching factor. Each pair of successive waypoints along a path can also be used as (start,goal) pairs for a planning task in another domain, thus reducing the search depth. Both these methods are used to focus and accelerate searches in complex domains, providing real-time efficiency without compromising on control fidelity. Section 9.5 describes the relationships between domains.

9.2   PLANNING DOMAINS

A problem domain is defined as $\Sigma = \langle \mathbf{S}, \mathbf{A}, c(s, s_n), h(s, s_{goal}) \rangle$, where the state space $\mathbb{S} = \{\mathbb{S}_{self} \times \mathbb{S}_{env} \times \mathbb{S}_{agents}\}$ includes the internal state of the agent $\mathbb{S}_{self}$, the representation of the environment $\mathbb{S}_{env}$, and other agents $\mathbb{S}_{agents}$. $\mathbb{S}_{self}$ may be modeled as a simple particle with a collision radius. $\mathbb{S}_{env}$ can be an environment triangulation with only static information or a uniform grid representation with dynamic obstacles. $\mathbb{S}_{agents}$ is defined by the vicinity within which neighboring agents are considered. Imminent threats may be considered individually or just represented as a density distribution at far-away distances. The action space $\mathbf{A}$ defines the set of all possible successors $\texttt{succ}(s)$ and predecessors $\texttt{pred}(s)$ at each state s, as shown in Equation 9.1:

$$\texttt{succ}(s) = \{s + \delta(s, i) | \Phi(s, s_n) = \text{TRUE} \ \forall i\}. \tag{9.1}$$

Here, $\delta(s, i)$ describes the ith transition, and $\Phi(s, s_n)$ is used to check if the transition from s to s_n is possible. The cost function $c(s, s_n)$ defines the cost of transition from s to s_n. The heuristic function $h(s, s_{goal})$ defines the estimate cost of reaching a goal state.

A problem definition $P = \langle \Sigma, s_{start}, s_{goal} \rangle$ describes the initial configuration of the agent, the environment, and other agents, along with the desired goal configuration in a particular domain. Given a problem definition $\mathbf{P}$ for domain Σ, a planner searches for a sequence of transitions to generate a space-time plan $\Pi(\Sigma, s_{start}, s_{goal}) = \{s_i | s_i \in \mathbf{S}(\Sigma)\}$ that takes an agent from s_{start} to s_{goal}.

9.3   MULTIPLE DOMAINS OF CONTROL

We define four domains which provide a suitable balance between global static navigation and fine-grained space-time control of agents in dynamic environments. Figure 9.2 illustrates the different domain representations for a given environment.

9.3.1   STATIC NAVIGATION MESH DOMAIN

The static navigation mesh domain, Σ_1, uses a triangulated representation of free space and only considers static immovable geometry. Dynamic obstacles and agents are not considered in this domain. The agent is modeled as a point mass, and valid transitions are between connected free spaces, represented as polygons. The cost function is the straight-line distance between the center points of two free spaces. Additional connections are also precomputed (or manually annotated) to represent transitions such as jumping with a higher defined cost. The heuristic function is the

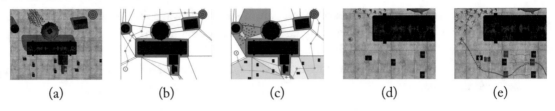

(a)   (b)   (c)   (d)   (e)

Figure 9.2: (a) Problem definition with initial configuration of agent and environment. (b) Global plan in static navigation mesh domain Σ_1 accounting for only static geometry. (c) Global plan in dynamic navigation mesh domain Σ_2 accounting for cumulative effect of dynamic objects. (d) Grid plan in Σ_3. (e) Space-time plan in Σ_4 that avoids dynamic threats and other agents.

Euclidean distance between a state and the goal. Searching for an optimal solution in this domain is very efficient and quickly provides a global path for the agent to navigate.

9.3.2   DYNAMIC NAVIGATION MESH DOMAIN

The dynamic navigation mesh domain, Σ_2, also uses triangulations to represent free spaces and coarsely accounts for dynamic properties of the environment to make a more informed decision at the global planning layer. We define a time-varying density field $\phi(t)$ which stores the density of movable objects (agents and obstacles) for each polygon in the triangulation at some point of time t.

9.3.3   GRID DOMAIN

The grid domain, Σ_3, discretizes the environment into grid cells where a valid transition is considered between adjacent cells that are free (diagonal movement is allowed). An agent is modeled as a point with a radius—its orientation and speed is not considered in this domain. This domain only accounts for the current position of dynamic obstacles and agents, and cannot predict collisions in space-time. The cost and heuristic are functions that measure the Euclidean distance between grid cells.

9.3.4   SPACE-TIME DOMAIN

The space-time domain, Σ_4, models the current state of an agent as a space-time position with a current velocity $(\mathbf{x}, \mathbf{v}, t)$. Σ_4 accounts for all obstacles (static and dynamic) and other agents. The traversability of a grid cell is queried in space-time by checking to see if movable obstacles and agents occupy that cell at that particular point of time, by using their published paths. For space-time collision checks, only agents and obstacles are considered that are within a certain region from the agent, defined using a foveal angle intersection. The cost and heuristic definitions have a great impact on the performance in Σ_4. We use an energy-based cost formulation that penalizes change in velocity with a non-zero cost for zero velocity. Jump transitions incur a higher cost.

The heuristic function penalizes states that are far away from s_{goal} in both space and time. This is achieved using a weighted combination of a distance metric and a penalty for a deviation of the current speed from the speed estimate required to reach s_{goal}.

9.4 PROBLEM DECOMPOSITION AND MULTI-DOMAIN PLANNING

Figure 9.3(a) illustrates the use of tunnels to connect each of the 4 domains, ensuring that a complete path from the agents initial position to its global target is computed at all levels. Figure 9.3(b) shows how Σ_2 and Σ_3 are connected by using successive waypoints in $\Pi(\Sigma_2)$ as start and goal for independent planning tasks in Σ_3. This relation between Σ_2 and Σ_3 allows finer-resolution plans to be computed between waypoints in an independent fashion. Limiting Σ_3 (and Σ_4) to plan between waypoints instead of the global problem instance insures that the search horizon in these domains is never too large, and that fine-grained space-time trajectories to the initial waypoints are computed quickly. However, completeness and optimality guarantees are relaxed as Σ_3 and Σ_4 never compute a single path to the global target.

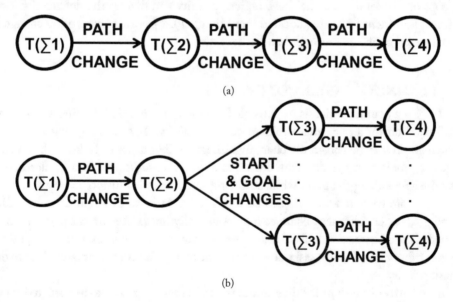

Figure 9.3: Relationship between domains. (a) Use of tunnels to connect each of the four domains. (b) Use of successive waypoints in $\Pi(\Sigma_2)$ as start, goal pairs to instantiate multiple planning tasks in Σ_3 and Σ_4.

We briefly describe the different events that are sent between planning tasks to trigger plan refinement and updates for the domain relationship in Figure 9.3(b). Σ_1 is first used to compute a path from s_{start} to s_{goal}, ignoring dynamic obstacles and other agents. $\Pi(\Sigma_1)$ is used to accelerate computations in Σ_2, which refines the global path to factor in the distribution of dynamic objects in the environment. Depending on the relationship between Σ_2 and Σ_3, a single planning task or multiple independent planning tasks are used in Σ_3. Finally, the plan(s) of $T(\Sigma_3)$ are used to accelerate searches in Σ_4.

Changes in s_{start} and s_{goal} trigger plan updates in $T(\Sigma_1)$, which are propagated through the task dependency chain. $T(\Sigma_2)$ monitors plan changes in $T(\Sigma_1)$ as well as the cumulative effect of changes in the environment to refine its path. Each $T(\Sigma_3)$ instance monitors changes in the waypoints along $\Pi(\Sigma_2)$ to repair its solution, as well as nearby changes in obstacle and agent position. Finally, $T(\Sigma_4)$ monitors plan changes in $T(\Sigma_3)$ (which it depends on) and repairs its solution to compute a space-time trajectory that avoids collisions with static and dynamic obstacles and other agents.

Events are triggered (outgoing edges) and monitored (incoming edges) by tasks, creating a cyclic dependency between tasks, with T_0 (agent execution) monitoring changes in the plan produced by the particular $T(\Sigma_4)$, which monitors the agent's most imminent global waypoint. Tasks that directly affect the agent's next decision, and tasks with currently invalid or sub-optimal solutions are given higher priority. Given the maximum amount of time to deliberate t_{max}, the agent pops one or more tasks that have highest priority and divides the deliberation time across tasks (most imminent tasks are allocated more time). Task priorities constantly change based on events triggered by the environment and other tasks.

9.4.1 PLANNING TASKS AND EVENTS

A task $T(\mathbf{P})$ is a planner which is responsible for generating and maintaining a valid (and ideally optimal) solution for a particular problem definition $\mathbf{P} = \langle \Sigma, s_{\text{start}}, s_{\text{goal}} \rangle$ where s_{start}, s_{goal}, and the search graph may be constantly changing. There are four types of tasks, each of which solves a particular problem in the domains described above. An additional task T_0 is responsible for moving the agent along the path, while enforcing steering and collision constraints.

Events are triggered and monitored by planning tasks in different domains. Changes in start and goal, or environment changes may potentially invalidate current plans, requiring plan refinement. Tasks that use tunnels to accelerate searches in more complex domains, monitor plan changes in other tasks. Finally, tasks observe the optimality status of their own plans to determine their task priority.

The priority of a task $p(T_a)$ determines the tasks that are picked to be executed at every time step, with tasks having smallest $p(T_a)$ chosen for execution ($p(T_a)$ is short for $p(T(\Sigma_a))$). Task T_0, which handles agent movement always has a priority of 1. Priority of other tasks is calculated as follows:

$$p(T_a) = \begin{cases} 1 \text{ if } T_a = T_0, \\ \mu(T_a, T_0) \cdot \Omega(T_a) \text{ otherwise,} \end{cases} \qquad (9.2)$$

where $\mu(T_a, T_0)$ is the number of edge traversals required to reach T_0 from T_a in the task dependency chain. $\Omega(T_a)$ denotes the current state of the plan of T_a and is defined as follows:

$$\Omega(T_a) = \begin{cases} 1 & \text{if SOLUTION\_INVALID} \\ \epsilon & \text{if plan inflation factor, } \epsilon > 1 \\ \infty & \text{if plan inflation factor, } \epsilon = 1, \end{cases} \tag{9.3}$$

where ϵ is the inflation factor used to determine the optimality bounds of the current plan for that task. The agent pops one or more tasks that have highest priority and divides the deliberation time available across tasks, with execution-critical tasks receiving more time. Tasks that have the same priority are ordered based on task dependency. Hence, T_0 is always executed at the end of every update after all planning tasks have completed.

The overall framework enforces strict time constraints. Given an allocated time to deliberate for each agent (computed based on desired frame rate and number of agents), the time resource is distributed based on task priority. In the remote event that there is no action to execute, the agent remains stationary (no impact on frame-rate) for a few frames (fraction of a second) until a valid plan is computed.

9.5 RELATIONSHIP BETWEEN DOMAINS

The complexity of the planning problem increases exponentially with increase in dimensionality of the search space—making the use of high-dimensional domains nearly prohibitive for real-time applications. In order to make this problem tractable, planning tasks must efficiently use plans in one domain to focus and accelerate searches in more complex domains. Section 9.5.1 describes a method for mapping a state from a low-dimensional domain to one or more states in a higher dimensional domain. The remainder of this section describes two ways in which plans in one domain can be used to focus and accelerate searches in another domain.

9.5.1 DOMAIN MAPPING

We define a $1 : n$ function $\lambda(s, \Sigma, \Sigma')$ that allows maps states in $S(\Sigma)$ to one or more equivalent states in $S(\Sigma')$:

$$\lambda(s, \Sigma, \Sigma') : s \to \{s_n | s_n \in S(\Sigma') \wedge s \equiv s_n\}. \tag{9.4}$$

The mapping functions are defined specifically for each domain pair. For example, $\lambda(s, \Sigma_1, \Sigma_2)$ maps a polygon $s \in S(\Sigma_1)$ to one or more polygons $\{s_n | s_n \in S(\Sigma_2)\}$ such that s_n is spatially contained in s. If the same triangulation is used for both Σ_1 and Σ_2, then there exists a one-to-one mapping between states. Similarly, $\lambda(s, \Sigma_2, \Sigma_3)$ maps a polygon $s \in S(\Sigma_2)$ to multiple grid cells $\{s_n | s_n \in S(\Sigma_3)\}$ such that s_n is spatially contained in s. $\lambda(s, \Sigma_3, \Sigma_4)$ is defined as follows:

$$\lambda(s, \Sigma_3, \Sigma_4) : (\mathbf{x}) \rightarrow \{(\mathbf{x} + W(\Delta \mathbf{x}), t + W(\Delta t))\}, \tag{9.5}$$

where $W(\Delta)$ is a window function in the range $[-\Delta, +\Delta]$. The choice of t is important in mapping Σ_3 to Σ_4. Since we use λ to effectively map a plan $\Pi(\Sigma_3, s_{start}, s_{goal})$ in Σ_3 to a tunnel in Σ_4, we can exploit the path and the temporal constraints of s_{start} and s_{goal} to define t for all states along the path. The total path length and the time to reach s_{goal} are calculated. This yields the approximate time of reaching a state along the path, assuming the agent is traveling at a constant speed.

9.5.2 MAPPING SUCCESSIVE WAYPOINTS TO INDEPENDENT PLANNING TASKS

Successive waypoints along the plan from one domain can be used as start and goal for a planning task in another domain. This effectively decomposes a planning problem into multiple independent planning tasks, each with a significantly smaller search depth.

Consider a path $\Pi(\Sigma_2) = \{s_i | s_i \in \mathbf{S}(\Sigma_2), \forall i \in (0, n)\}$ of length n. For each successive waypoint pair (s_i, s_{i+1}), we define a planning problem $\mathbf{P}_i = \langle \Sigma_3, s_{start}, s_{goal} \rangle$ such that $s_{start} = \lambda(s_i, \Sigma_2, \Sigma_3)$ and $s_{goal} = \lambda(s_{i+1}, \Sigma_2, \Sigma_3)$. Even though λ may return multiple equivalent states, we choose only one candidate state. For each problem definition $\mathbf{P}_i$, we instantiate an independent planning task $T(\mathbf{P}_i)$ which computes and maintains path from s_i to s_{i+1} in Σ_3. Figure 9.3 illustrates this connection between Σ_2 and Σ_3.

9.6   RESULTS

9.6.1   COMPARATIVE EVALUATION OF DOMAINS

The comparative evaluations of domains shows that no single domain can efficiently solve the challenging problem instances that were sampled. The use of tunnels significantly reduces the effective branching factor of the search in Σ_3 and Σ_4, while mapping successive waypoints in $\Pi(\Sigma_2)$ to multiple independent planning tasks reduce the depth of the search in Σ_3 and Σ_4, without significantly impacting success rate and quality. For the remaining results, we adopt this domain relationship as it works well for our application of simulating multiple goal-directed agents in dynamic environments at interactive rates. Users may choose a different relationship based on their specific needs.

9.6.2   PERFORMANCE

We measure the performance of the framework by monitoring the execution time of each task type, with multiple instances of planning tasks for Σ_3 and Σ_4. We limit the maximum deliberation time $t_{max} = 10$ ms, which means that the total time executing any of the tasks at each frame cannot exceed 10 ms. For this experiment, we limit the total number of tasks that can be executed in a single frame to 2 (including T_0) to visualize the execution time of each task over

 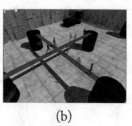 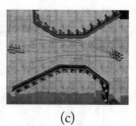 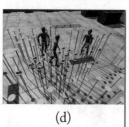

| (a) | (b) | (c) | (d) |

Figure 9.4: Different scenarios. (a) Agents crossing a highway with fast moving vehicles in both directions. (b) Four agents solving a deadlock situation at a 4-way intersection. (c) Twenty agents distributing themselves evenly in a narrow passage, to form lanes both in directions. (d) A complex environment requiring careful foot placement to obtain a solution.

different frames. Figure 9.5 illustrates the task execution times of a single agent over a 30 second simulation for the scenario shown in Figure 9.2(a). The execution task T_0, which is responsible for character animation and simple steering, takes approximately 0.4–0.5 ms of execution time every frame. Spikes in the execution time correlate to events in the world. For example, a local non-deterministic change in the environment (frames 31,157) triggers a plan update in $T(\Sigma_3)$, which in turn triggers an update in $T(\Sigma_4)$. A global change such as a crowd blocking a passage or a change in goal (frames 39, 237,281) triggers an update in $T(\Sigma_2)$ or $T(\Sigma_1)$ which in turn propagates events down the task dependency chain.

Note that there are often instances during the simulation when the start or goal changes significantly or when plans are invalidated, requiring planning from scratch. However, we ensure that our framework meets real-time constraints due to the following design decisions: (a) limiting the maximum amount of time to deliberate for the planning tasks, (b) intelligently distributing the available computational resources between tasks with highest priority, and (c) increasing the inflation factor to quickly produce a sub-optimal solution when a plan is invalidated, and refining the plan in successive frames.

Memory

$T(\Sigma_1)$ and $T(\Sigma_2)$ precompute navigation meshes for the environment whose size depend on environment complexity, but are shared by all agents in the simulation. The runtime memory requirement of these tasks is negligible since it expands very few nodes. The memory footprint of $T(\Sigma_3)$ and $T(\Sigma_4)$ is defined by the number of nodes visited by the planning task during the course of a simulation. Since each planning task in Σ_3 and Σ_4 searches between successive waypoints in the global plan, the search horizon of the planners is never too large. On average, the number of visited nodes is 75 and 350 for $T(\Sigma_3)$ and $T(\Sigma_4)$, respectively, with each node occupying 16–24 bytes in memory. For 5 running instances of $T(\Sigma_3)$ and $T(\Sigma_4)$, this amounts to approximately 45 KB of memory per agent. Additional memory for storing other plan containers such as OPEN

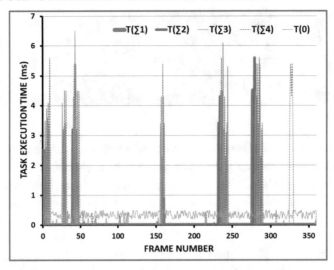

Figure 9.5: Task execution times of the different tasks in our framework over the course of a 60 second simulation.

and CLOSED are not considered in this calculation as they store only node references and are cleared after every plan iteration.

Scalability

Our approach scales linearly with increased number of agents. The maximum deliberation time *for all* agents can be chosen based on the desired framerate which is then distributed among agents and their respective planning tasks at each frame. The cost of planning is amortized over several frames and all agents need not plan simultaneously. Once an agent computes an initial plan, it can execute the plan with efficient update operations until it is allocated more deliberation time. If its most imminent plan is invalidated, it is prioritized over other agents and remains stationary until computational resources are available. This ensures that the simulation meets the desired framerate.

9.6.3   SCENARIOS

We demonstrate the benefits of our framework by solving many challenging scenarios (see Figure 9.4) requiring space-time precision, explicit coordination between interacting agents, and the factoring of dynamic information (obstacles, moving platforms, user-triggered changes, and other agents) at all stages of the decision process. All results shown here were generated at 30 fps or higher, which includes rendering and character animation. We use an extended version of the ADAPT character animation system [Shoulson et al., 2014a] for the results described below. We also carried out a more comprehensive analysis of our framework using the benchmarks described in [Kapadia et al., 2011c, Singh et al., 2009b]. Interested readers may refer to [Kapadia et al., 2013] for more details.

Deadlocks

Multiple oncoming and crossing agents in narrow passageways cooperate with each other with space-time precision to prevent potential deadlocks. Agents observe the presence of dynamic entities at waypoints along their global path and refine their plan if they notice potentially blocked passageways or other high cost situations. Other crowd simulators often deadlock for these scenarios, while a space-time planner by itself does not scale well for many agents.

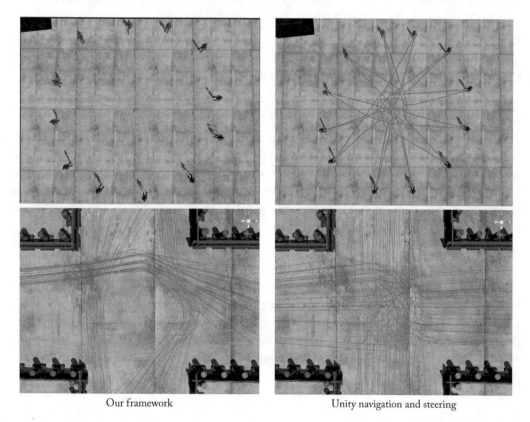

Our framework   Unity navigation and steering

Figure 9.6: Trajectory comparison of our method with an off-the-shelf predictive steering algorithm in the Unity game engine. Our framework minimizes deviation and uses speed variations to avoid collisions in space-time.

Choke Points

This scenario shows our approach handling agents arriving at a common meeting point at the same time, producing collision-free straight trajectories. Figure 9.6 compares the trajectories produced using our method with an off-the-shelf navigation and predictive collision avoidance algorithm in the Unity game engine. Our framework produces considerably smoother trajectories and minimizes deviation by using subtle speed variations to avoid collisions in space-time.

Unpredictable Environment Change

Our method efficiently repairs solutions in the presence of unpredictable world events, such as the user-placement of obstacles or other agents, which may invalidate current paths.

Road Crossing

The road crossing scenario demonstrates 40 agents using space-time planning to avoid fast moving vehicles and other crossing agents.

Lane Selection for Bi-directional Traffic

This scenario requires agents to make a navigation decision in choosing one of 4 lanes created by the dividers. Agents distribute themselves among the lanes, while bi-directional traffic chooses different lanes to avoid deadlocks. This scenario requires non-deterministic dynamic information (other agents) to be accounted for while making global navigation decisions. This is different from emergent lane formation in crowd approaches, which bottlenecks at the lanes and cause deadlocks without a more robust navigation technique.

Four-way Crossing

We simulate 100 oncoming and crossing agents in a four-way crossing. The initial global plans in Σ_1 take the minimum distance path through the center of the crossing. However, Σ_2 predicts a space-time collision between groups at the center and performs plan refinement so that agents deviate from their optimal trajectories to minimize group interactions. A predictive steering algorithm only accounts for imminent neighboring threats and is unable to avoid mingling with the other groups (second row of Figure 9.6).

Space-Time Goals

We demonstrate a complex scenario where 4 agents in focus (additional agents are also simulated) have a temporal goal constraint, defined as an interval ($40 + / - 1$ s). Agents exhibit space-time precision while jumping across moving planes to reach their target and the temporal goal significantly impacts the decision making at all levels. The space-time domain alone may be unable to meet the temporal constraint and require plans to be modified in earlier domains. No other approach can solve this with real-time constraints.

Many of these scenarios *cannot* be solved by the current state of the art in multi-agent motion planning, which is able to either handle a single agent with great precision, or simulate many simple agents that exhibit only reactive collision avoidance.

CHAPTER 10

Motion Planning for Character Motion Synthesis

The topics discussed in the previous chapters have covered a variety of methods for addressing path planning and navigation problems in virtual worlds. Going beyond the typical applications of motion planning, there are a number of planning techniques which have been developed over the past years for addressing problems related to full-body motion synthesis for virtual characters. This chapter presents an overview of representative approaches in this area, with focus on data-based motion planning for synthesizing locomotion, and on sampling-based manipulation planning for virtual characters.

In general, the presented methods employ motion planners in order to produce complex full-body motions in cluttered environments, bringing virtual characters to higher levels of motion autonomy. The general drawback of the presented approaches is that, although they can be employed in interactive applications, they are usually too expensive for highly interactive scenes or for controlling multiple characters at the same time. Therefore, these methods are not yet popular in practical applications. However, the growing demand for intelligent virtual characters for a variety of serious applications, such as in training, education and therapy, justify the quest for fully autonomous interactive virtual characters. Several of the approaches and challenges are also motivated and shared by applications in humanoid robotics.

10.1 PLANNING LOCOMOTION FROM MOTION CAPTURE DATA

The most popular approach for locomotion synthesis around obstacles is to reduce the planning problem to 2D path planning and to rely on a locomotion controller that is able to follow 2D paths. Realistic results are often achieved with locomotion controllers built based on motion capture clips carefully organized and parameterized [Gleicher et al., 2003, Park et al., 2004]. Once a suitable locomotion controller is built, the approach achieves fast results and is commonly employed in real-time applications and video games.

Relying on locomotion controllers, however, still requires significant engineering work for achieving generic and realistic results, and several planning methods have been developed with the goal to automatically synthesize locomotion directly from an input set of unorganized motion capture data. Such an approach largely simplifies the goal of transferring to the virtual character example locomotion performances with different styles captured directly from real actors. While locomotion controllers still represent the best approach in terms of computation performance, planning algorithms operating on motion capture data can produce better-quality results by imposing minimal modification to the original quality of the data, they can plan motions of higher complexity, and they can also be implemented to run at interactive frame rates.

There is an extensive body of research on planning motions directly from unstructured motion capture clips based on the concept of *motion graphs* [Arikan and Forsyth, 2002, Kovar et al., 2002]. Unstructured graphs are automatically built from input clips by creating links between frames in the data that are close enough to each other, according to a given error metric threshold. Discrete graph search can be then employed to find a motion in the graph that satisfies given constraints. Motions can be generated for reaching a target location, or for remaining as close as possible to a 2D path to be followed. Locomotion solutions are concatenations of motion data pieces, according to the transitions in the motion graph. The search process results in an *unrolling* of the graph in the environment, and because minimal concatenation blending is applied between transitions, highly realistic locomotion is achieved.

The drawback of searching in a motion graph is that the process can easily become too expensive for achieving real-time performances, and therefore recent research has focused on improving the performance of the overall approach. A first technique for improving performance is to prune the search branches that lie far away from a given 2D path with clearance from obstacles [Mahmudi and Kallmann, 2013], as illustrated in Figure 10.1. The approach also relies on graph link creation at specific locomotion transition points that improve graph connectivity.

In addition, the graph search can be highly optimized by relying on pre-computation of motion maps built from the motion graph. Motion maps encode the graph unrolling process for each vertex up to a given concatenation depth. Later at run-time the relevant motion maps are retrieved and their branches that are closest to an input path are efficiently detected and concatenated to achieve path following at interactive rates [Mahmudi and Kallmann, 2012]; see Figure 10.2 for examples.

The approach illustrates how advanced planning techniques can be employed in order to efficiently achieve high-quality results from unstructured motion capture data. The motion graph approach, however, does not extend naturally to reaching or manipulation problems, which represent types of motions that do not have cycles and that often require fine-level planning around obstacles. Instead, motion synthesis for object manipulation can be naturally translated into a high-dimensional motion planning problem, and the following section summarizes popular approaches in the area.

10.2   PLANNING MOTIONS IN HIGH DIMENSIONS

Let A denote a structure to be controlled. A motion planning problem for A involves computing a collision-free motion connecting an initial configuration of A to a goal configuration of A. A configuration of A is a vector encoding all the degrees of freedom of A, such that it denotes a

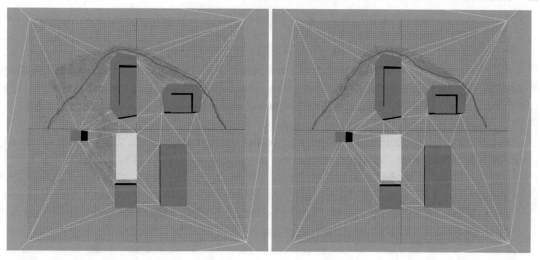

Figure 10.1: The left image illustrates the space traversed by a standard search over a motion graph in order to reach a destination point while avoiding obstacles. The right image shows how the search space is reduced when pruning the branches of the search to a channel around a pre-computed 2D path to the goal.

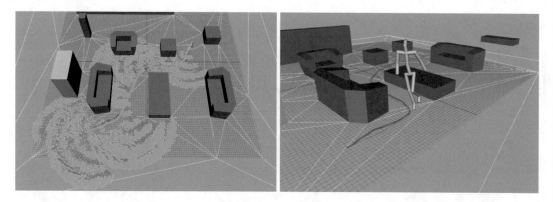

Figure 10.2: The left image shows four motion maps used in a path following query. Motion maps are search tree expansions efficiently pre-computed and stored, and then employed at run-time for achieving high-quality locomotion at interactive rates. The right image illustrates a result with locomotion data in happy style.

complete pose specification of A. For example, if A is a generic solid shape that can translate and rotate in the 3D space, a configuration of A will be specified with 6 parameters: 3 positional values and 3 values representing its orientation. If A is an articulated structure, all articulation parameters such as the joint angles needed to specify a pose of A are also part of the configuration description.

For instance, $\mathcal{A}$ may represent an arm of a virtual character and some manipulation problems can be reduced to computing motions for $\mathcal{A}$.

Given the high-dimensionality of the configuration spaces involved in manipulation problems, planners that are able to operate in these spaces have to rely on operations that are simple enough to be implemented in high dimensions. This basically rules out approaches based on exact cell decomposition and the most successful generic methods are *sampling-based* methods.

Sampling-based planners rely on configuration space sampling to build a graph or tree representation in the collision-free portion of the configuration space. Probablistic Roadmaps (PRMs) [Kavraki et al., 1996] represent one of the first sampling-based methods that became highly successful in solving practical problems. Given an environment, randomly sampled points in the configuration space of the problem are sampled in order to construct a roadmap graph which is later used to answer motion planning queries. The roadmap construction is performed as a pre-computation phase. Several configurations are generated by random sampling and those that are determined to be valid and collision-free are stored and will become nodes of the roadmap. Usually the sampling process is executed until a specified number of valid nodes is reached. Then, each node is connected by an edge to its nearby nodes if the edge connection represents a valid and collision-free motion between the two nodes. Figure 10.3 illustrates the overall process.

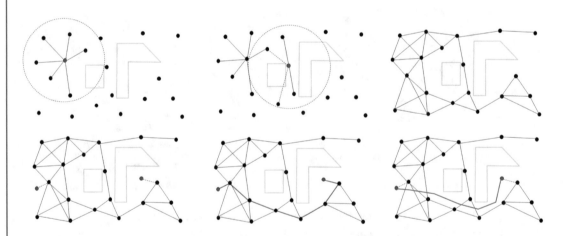

Figure 10.3: PRMs are built from a number of valid configuration samples. Each sample is connected to a nearby sample if a valid motion can be computed between the two configurations being connected (top-left). Non-valid connections are discarded (top-center) and the result is a graph in the free space (top-right). At run-time query points are connected to their nearest nodes in the graph (bottom-left) and graph search is executed to find a connection between the two points (bottom-center). After a solution is found, smoothing operations are often needed to improve the solution quality (bottom-right).

The example in Figure 10.3 can be seen as a waypoint graph similar to the examples illustrated in Chapter 1, however, the steps involved allow PRMs to be applied to planning problems in any dimension. Given that roadmaps can be re-used for multiple queries, PRMs are categorized as a multi-query method. Unfortunately, the cost of pre-computing roadmaps can be high and a roadmap is only valid while the environment does not change. Whenever these limitations become important single-query methods are preferred.

Single-query methods do not employ pre-computation and instead they build a structure in the free space specifically optimized for each motion planning query being solved. One of the most popular single-query method is the *Rapidly-exploring Random Tree* (RRT) [Lavalle, 1998]. In this case the structure built during the search is a tree, and once the algorithm successfully finishes, a solution motion is naturally achieved as a branch of the tree. Figure 10.4 illustrates the tree growing process. The process will stop when a branch of the tree reaches a target configuration.

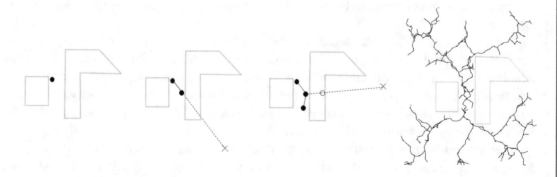

Figure 10.4:   In left-right order: RRTs are built starting from a tree initialized at a given valid configuration. Configurations are progressively sampled and the closest node in the tree is extended toward each sample, with invalid extensions being rejected. The result is a tree that grows covering the free space of the problem.

PRMs and RRTs represent two basic approaches in the area of sampling-based motion planning and several variations and additional methods are also available. Additional information can be found on textbooks in the area [Choset et al., 2005, Latombe, 1990, LaValle, 2006]. More recently, these algorithms have been extended to converge toward optimal solutions as the sampling progresses [Karaman and Frazzoli, 2011], and a large variety of additional extensions can also be found in the research literature. The described PRM and RRT approaches represent the two most popular baseline methods for generic motion planning problems, and they are also often used as the underlying technique of more advanced approaches.

One drawback of sampling-based planners is the required computation time, which can become significant in difficult problems. In addition, it is difficult to determine when the search has sufficiently explored the free space. These methods are said to be probabilistic complete, which means that only if the random sampling can indefinitely continue that a solution will be eventually found, if one exists.

The following section discusses the application of sampling-based planning to solve reaching and manipulation problems.

10.3 PLANNING MANIPULATION MOTIONS

There is a clear trade-off between employing multi-query or single-query methods for solving object manipulations. The very fact that the object being manipulated will change its position means that multi-query roadmaps have to be re-built or adapted every time there is a change in the environment. Approaches for maintaining dynamic roadmaps were described in [Kallmann and Matarić, 2004] and can speed up computation performance in several situations; however, such methods are only effective up to a certain level of problem complexity. After a certain point roadmap maintenance operations become too expensive and single-query methods become more effective.

Given the difficulty in maintaining a roadmap structure in a way that is effective for any situation, the typical generic approach for planning object reaching and manipulation is to employ a bidirectional single-query planner. In the RRT case, a bidirectional implementation will grow two trees at the same time, one rooted at the initial configuration, and the other rooted at the goal configuration, until they connect and generate a solution [Kuffner and LaValle, 2000]. An implementation is presented in Algorithm 24.

Algorithm 24 can be directly applied to a reaching or manipulation problem. In its simplest form, a configuration q will contain all the joint angles of the arm of the character. Additional joint angles for the spine can also be considered in order to achieve a larger reachable space. The planning problem is then to compute a motion between an initial configuration s and a goal configuration t. Usually, s is the current posture of the character and t is computed by Inverse Kinematics from a target location and orientation for the hand to reach. In the case of an object manipulation, it is assumed that the object is already grasped and attached to the hand of the character, such that the problem is reduced to planning an arm motion while holding the object.

Line 5 of Algorithm 24 requires a configuration sampling routine. This routine guides the whole search and therefore is of main importance. The basic process samples random valid joint values inside their range limits, and a sample is only returned if it has no collisions; otherwise additional samples are generated until a collision-free one is found. In typical cases randomly sampling postures has the effect of biasing the search toward large volumes of free space located at the sides of the character. Since realistic manipulations are mostly expected to remain in front of the character, a possible simple correction technique is to bias the elbow flexion sampling to almost full flexion. This has the effect of avoiding solutions with the arm outstretched, resulting in more natural motions. As the algorithm performs a bidirectional search, it also contributes to decomposing the manipulation in two natural phases: bringing the arm closer to the body and then extending it toward the goal. Biasing the wrist joint value to a small range is also beneficial because

---

Algorithm 24 - Bidirectional RRT Planning Algorithm

Input: source configuration s and target configuration t.

Output: valid motion from s to t, or failure.

1: **BidirectionalRRT**(s, t)
2:   $T_1 \leftarrow$ search tree initialized with s as root;
3:   $T_2 \leftarrow$ search tree initialized with t as root;
4:   **while** (elapsed time $\leq$ maximum allowed time) **do**
5:     $s_{sample} \leftarrow$ SAMPLEVALIDCONFIGURATIONPOSTURE();
6:     $s_1 \leftarrow$ closest node to s_{sample} in T_1;
7:     $s_2 \leftarrow$ closest node to s_{sample} in T_2;
8:     **if** (INTERPOLATIONVALID(s_1, s_2)) **then**
9:       **return** MAKEPATH$(root(T_1), s_1, s_2, root(T_2))$;
10:    **end if**
11:    $s_{exp} \leftarrow$ NODEEXPANSION $(s_1, s_{sample}, \epsilon)$;
12:    **if** ($s_{exp} \neq null$ **and** INTERPOLATIONVALID(s_{exp}, s_2)) **then**
13:      **return** MAKEPATH$(root(T_1), s_{exp}, s_2, root(T_2))$;
14:    **end if**
15:    Swap T_1 and T_2;
16:   **end while**
17: **return** failure;

---

the wrist is expected to only be secondarily used for avoiding obstacles. A number of additional heuristics can be included to improve the quality of the results [Kallmann, 2005, Kallmann et al., 2003].

Lines 6 and 7 require searching for the closest configuration in each tree. A common distance metric is to take the maximum of the Euclidian distances between the position of corresponding joints, posed at each configuration. Lines 8 and 12 check if the interpolation between two configurations is valid, what can be performed with several discrete collision checks along the interpolation or with efficient continuous methods [Schwarzer et al., 2002].

Factor ε gives the incremental step taken during the search. Large steps make trees to grow quickly but with more difficulty to capture the free configuration space around obstacles. Too small values generate roadmaps with too many nodes, slowing down the tree expansion. The expanded node s_{exp} in line 11 uses s_{sample} as growing direction and it can be determined by interpolation using an interpolation factor such as $t = \varepsilon/d$, where $d = distance(s_1, s_{sample})$.

A final post-processing step for smoothing the path is usually required. The simplest approach is to apply a few random linearization iterations. Each linearization consists of selecting two random configurations (not necessarily nodes) along the solution path and replacing the subpath between them by a straight interpolation if the replacement is a valid path [Schwarzer et al., 2002]. Figure 10.5 illustrates a manipulation computed with Algorithm 24.

Figure 10.5: The top-left image shows the RRT bidirectional trees expanding toward free spaces. The remaining images show frames of the smoothed solution motion. The red trajectory depicts the trajectory of the wrist joint along the solution. The bottom images show corresponding frames from a different point of view.

Among the several extensions that are possible, one approach for improving the efficiency of a single-query sampling-based planner is to bias the search using some type of learning based on previous experiences. For instance, an algorithm based on attractor points learned from previous executions has been proposed to guide the planning in subsequent queries in order to improve the overall planning performance [Jiang and Kallmann, 2007]. Among several other contributions in the literature, experience-based enhancements to PRMs have also been proposed [Coleman et al., 2015]. Motion planning is an active research area in robotics and several additional methods are available. Implementation efforts gathering established methods in a unified platform are also available [Şucan et al., 2012].

10.4 INTEGRATED PLANNING OF LOCOMOTION AND OBJECT MANIPULATION

The methods discussed in the previous sections present different approaches taken for addressing motion synthesis for locomotion and object manipulation. However, in most cases locomotion has to be planned in conjunction with manipulation. This makes sense not only for producing realistic results, but also because in many cases locomotion is essential for placing the character at a suitable location for action execution.

Based on these observations, approaches for planning locomotion integrated with generic upper-body actions are important to be developed. In robotics, planning a motion that integrates locomotion with upper-body actions is often referred to as a mobile manipulation problem. For virtual characters one important additional problem is to achieve realistic human-like solutions, which is often addressed by incorporating motion capture data into a given planning method. We now describe such an approach integrating all these elements, and which is based on the previous methods presented in this chapter. The approach achieves planning of human-like full-body motions in complex environments [Huang et al., 2011].

Given the initial pose q_i of the character, the method starts by unrolling a locomotion graph structure as presented in Section 10.1. Every time one branch of the locomotion graph expansion reaches a character pose q_a that is close enough to the action target, q_a is then considered as a transition point to the upper body action. Configuration q_a becomes the initial posture for the upper-body action planner, which will in turn launch a bidirectional search attempting to reach a goal posture in Q_g, which is the set of goal poses satisfying the action completion. For example, in the case of object grasping, Q_g represents valid ways of grasping the object. If the upper-body planner is not successful after a fixed number of iterations, the locomotion planner continues to expand toward additional candidate body placements until the action can be executed or until a maximum time limit is reached, in which case the overall planner returns failure. Figure 10.6 illustrates this bi-modal search procedure.

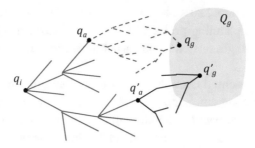

Figure 10.6: The locomotion planner unrolls motion capture locomotion clips (left blue tree), and for each candidate initial action posture, a bidirectional upper-body action planning is performed. Here q_a was not able to generate a solution after a fixed maximum number of iterations and a new body placement q'_a was finally successful. The overall approach interleaves upper-body action planning with the search for suitable body placements.

Every time one branch of the locomotion graph expansion reaches a character pose q_a that is close enough to the action target, q_a becomes a candidate initial pose for initializing the action planner. The upper-body action is based on a bidirectional RRT, however extended to operate on the blending space of a collection of similar and time-aligned example motions, which are realistic upper-body action instances collected from motion capture. The result is an upper-body planner able to address generic types of actions and to achieve realistic results.

The overall method quickly expands several body placements nearby the action target, providing to the upper-body planner several starting points for planning the upper-body action. The overall full-body planner therefore evaluates different body placements until a suitable valid and collision-free action is found. Figure 10.7 illustrates one solution motion computed with the method.

Figure 10.7: Example pouring motion computed among obstacles and integrated with locomotion. Locomotion is based on unrolling a feature-based motion graph and action planning relies on a bidirectional planner in the blending space of example actions [Huang et al., 2011].

The overall approach of decomposing a high-dimensional planning problem into planning at different modes or controlling multiple skills is related to the topic of multi-modal motion planning [Hauser and Latombe, 2010, Kallmann et al., 2010]. In general, multi-modal planning well addresses the typical planning problems faced by autonomous virtual characters in serious applications, even if challenges related to skill synchronization and speed of computation are still significant.

In summary, the presented examples demonstrate that motion planning represents a powerful approach for controlling autonomous characters in virtual worlds and several methods are available for addressing a number of constraints and requirements. As virtual worlds become more and more important in the next generation of applications, the importance of motion planning techniques for achieving truly autonomous virtual characters can only be expected to increase.

CHAPTER 11

Epilogue

This book covered a holistic presentation of foundational and advanced topics in discrete search, computational geometry, and motion planning techniques, including their state-of-the-art applications to animating autonomous virtual humans in interactive virtual worlds. We adopted a bottom-up perspective, systematically introducing the readers to the classical methods and to how they can be extended to meet the growing demands of today's interactive applications where addressing large and complex environments with real-time performances is of paramount importance. This book has targeted a trade-off between covering foundational topics that are often not taken into account by practitioners, and at the same time demonstrating their practical application to typical commercial settings such as in computer games and other types of virtual worlds. This is in contrast to other existing volumes on related areas which are often targeted at either academic researchers or industry practitioners, and thus not connecting both.

There are several recent contributions that have not been included in this volume. Recent work has explored how geometric approaches for environment analysis may be applied to annotate environments with contact semantics, in order to facilitate contact-rich character motions such as *Parkur* or *Freeruning* motions in complex environments [Kapadia et al., 2016]. Complementary to the problem of navigation and path-finding is the problem of *steering* agents along the planned paths while avoiding collisions with other agents and the environment. There is extensive literature on this topic, and we refer interested readers to other volumes that cover the area in greater depth [Kapadia et al., 2015b]. There is also a growing need to analyze and evaluate simulations of autonomous agents, in order to detect anomalies [Boatright et al., 2012, Kapadia et al., 2009a], and to meet various quality and performance criteria [Kapadia et al., 2011b,c, Singh et al., 2009b,b, 2008]. These quantitative metrics can also be used to automatically optimize the behavior of simulated agents in order to satisfy user-defined objectives [Berseth et al., 2013a, 2014b] and environments [Berseth et al., 2014a, 2013b, 2015].

Planning approaches to character motion synthesis also count on a number of related contributions which have not been discussed in this book. In particular, a multi-modal planning approach has been recently developed for planing motions for virtual demonstrators delivering information to observers at arbitrary locations, and according to positioning rules extracted from experimental studies with human subjects [Huang and Kallmann, 2015]. Additional multi-modal planning techniques for synthesizing whole-body human-like motions have also been recently developed based on motion capture data in order to achieve complex mobile-manipulations [Mahmudi and Kallmann, 2015]. Integration of motion capture data with planners is also an important area which requires new approaches for processing motion capture data, such as for quantifying deformation of locomotion data [Juarez-Perez et al., 2014], and for parameterizing motions captured by direct demonstrations [Camporesi et al., 2010]. More generally, planning may also be used to synthesize the behaviors of autonomous virtual humans [Kapadia et al., 2013e, 2011a] in an effort to convey compelling narratives [Kapadia et al., 2015a]. This topic is beyond the scope of this book and we refer the reader to related contributions on the area [Shoulson et al., 2013a,b, 2014a].

Finally, this book did not cover how planning agents may apply to other real-world problem domains such as large-scale evacuations [Helbing et al., 2000], disaster and security simulations [Helbing et al., 2007, Helbing and Mukerji, 2012], or visual effects in movies [Massive Software Inc., 2010]. Each of these domains exhibit their own unique set of challenging problems.

Bibliography

Amato, N. M., Goodrich, M. T., and Ramos, E. A. (2000). Linear-time triangulation of a simple polygon made easier via randomization. In *Proceedings of the 16th Annual ACM Symposium of Computational Geometry*, pages 201–212. DOI: 10.1145/336154.336206. 32

André, E., Bosch, G., Herzog, G., and Rist, T. (1986). Characterizing trajectories of moving objects using natural language path descriptions. In *Proceedings of the 7th European Conference on Artificial Intelligence*, pages 1–8. 65

Arikan, O. and Forsyth, D. A. (2002). Synthesizing constrained motions from examples. *Proceedings of SIGGRAPH*, 21(3):483–490. 154

Arkin, R. (1987). Motor schema based navigation for a mobile robot: An approach to programming by behavior. In *Proceedings of the 1987 IEEE International Conference on Robotics and Automation*, volume 4, pages 264–271. DOI: 10.1109/ROBOT.1987.1088037. 65

Barbehenn, M. and Hutchinson, S. (1995). Efficient search and hierarchical motion planning by dynamically maintaining single-source shortest paths trees. *Robotics and Automation, IEEE Transactions on*, 11(2):198–214. DOI: 10.1109/70.370502. 68

Beacco, A., Pelechano, N., Kapadia, M., and Badler, N. I. (2015). Footstep parameterized motion blending using barycentric coordinates. *Computers & Graphics*, 47:105–112. DOI: 10.1016/j.cag.2014.12.004. 138

Berg, J., Guy, S., Lin, M., and Manocha, D. (2011). Reciprocal n-body collision avoidance. In Pradalier, C., Siegwart, R., and Hirzinger, G., editors, *Robotics Research*, volume 70 of *Springer Tracts in Advanced Robotics*, pages 3–19. Springer Berlin Heidelberg. 119

Berseth, G., Haworth, M. B., Kapadia, M., and Faloutsos, P. (2014a). Characterizing and optimizing game level difficulty. In *Motion in Games*, pages 153–160, New York, NY, ACM. DOI: 10.1145/2668084.2668100. 163

Berseth, G., Kapadia, M., and Faloutsos, P. (2013a). Automated parameter tuning for steering algorithms. In *ACM SIGGRAPH/Eurographics Symp. on Computer Animation, Poster Proc.*, SCA '13, pages 115–124, New York, NY, ACM. 163

Berseth, G., Kapadia, M., and Faloutsos, P. (2013b). Steerplex: Estimating scenario complexity for simulated crowds. In *Motion on Games*, MIG '13, pages 45:67–45:76, New York, NY, ACM. DOI: 10.1145/2522628.2522650. 163

Berseth, G., Kapadia, M., Haworth, B., and Faloutsos, P. (2014b). SteerFit: Automated Parameter Fitting for Steering Algorithms. In *ACM SIGGRAPH/Eurographics Symp. on Computer Animation*, SCA '14, New York, NY, ACM. DOI: 10.2312/sca.20141129. 163

Berseth, G., Usman, M., Haworth, B., Kapadia, M., and Faloutsos, P. (2015). Environment optimization for crowd evacuation. *Computer Animation and Virtual Worlds*, 26(3-4):377–386. DOI: 10.1002/cav.1652. 163

Bhattacharya, P. and Gavrilova, M. (2008). Roadmap-based path planning - using the voronoi diagram for a clearance-based shortest path. *Robotics Automation Magazine, IEEE*, 15(2):58–66. DOI: 10.1109/MRA.2008.921540. 48, 59

Bhattacharya, S., Likhachev, M., and Kumar, V. (2012a). Search-based path planning with homotopy class constraints in 3d. In *Proceedings of the 26th AAAI Conference on Artificial Intelligence*. 65

Bhattacharya, S., Likhachev, M., and Kumar, V. (2012b). Topological constraints in search-based robot path planning. *Autonomous Robots*, 33(3):273–290. DOI: 10.1007/s10514-012-9304-1. 65

Biknevicius, A. and Reilly, S. (2006). Correlation of symmetrical gaits and whole body mechanics: debunking myths in locomotor biodynamics. *Journal of Experimental Zoology Part A: Comparative Experimental Biology*, 305A(11):923–934. DOI: 10.1002/jez.a.332. 134

Björnsson, Y. and Halldórsson, K. (2006). Improved heuristics for optimal path-finding on game maps. In Laird, J. E. and Schaeffer, J., editors, *Proceedings of the AAAI Conference on Artificial Intelligence and Interactive Digital Entertainment (AIIDE)*, pages 9–14. The AAAI Press. 27

Bleiweiss, A. (2009). Scalable multi agent simulation on the gpu. In *GPU Technology Conference*. 64

Boatright, C. D., Kapadia, M., and Badler, N. I. (2012). Pedestrian anomaly detection using context-sensitive crowd simulation. In *International Workshop on Pattern Recognition and Crowd Analysis*. 163

Botea, A., Müller, M., and Schaeffer, J. (2004). Near optimal hierarchical path-finding. *Journal of Game Development*, 1:7–28. 64

Buluc, A., Gilbert, J. R., and Budak, C. (2010). Solving Path Problems on the GPU. *Journal of Parallel Computing*, 36(5-6):241–253. DOI: 10.1016/j.parco.2009.12.002. 64

Camporesi, C., Huang, Y., and Kallmann, M. (2010). Interactive motion modeling and parameterization by direct demonstration. In *Proceedings of the 10th International Conference on Intelligent Virtual Agents (IVA)*. DOI: 10.1007/978-3-642-15892-6_9. 164

Camporesi, C. and Kallmann, M. (2014). Computing shortest path maps with gpu shaders. In *Proceedings of Motion in Games (MIG)*. DOI: 10.1145/2668064.2668092. 41

Chazelle, B. (1982). A theorem on polygon cutting with applications. In *SFCS '82: Proceedings of the 23rd Annual Symposium on Foundations of Computer Science*, pages 339–349. IEEE Computer Society. DOI: 10.1109/SFCS.1982.58. 32

Chazelle, B. (1987). Approximation and decomposition of shapes. In Schwartz, J. T. and Yap, C. K., editors, *Algorithmic and Geometric Aspects of Robotics*, pages 145–185. Lawrence Erlbaum Associates, Hillsdale, NJ. 5, 6

Chazelle, B. (1991). Triangulating a simple polygon in linear time. *Discrete Computational Geometry*, 6(5):485–524. DOI: 10.1007/BF02574703. 32

Chen, M., Chowdhury, R. A., Ramachandran, V., Roche, D. L., and Tong, L. (2007). Priority queues and Dijkstra's algorithm. Technical Report TR-07-54, UTCS Technical Report. 20

Chew, L. P. (1985). Planning the shortest path for a disc in $O(n^2 log n)$ time. In *Proceedings of the ACM Symposium on Computational Geometry*. DOI: 10.1145/323233.323261. 35

Chin, F., Snoeyink, J., and Wang, C. A. (1999). Finding the medial axis of a simple polygon in linear time. *Discrete and Computational Geometry, In ISAAC: 6th International Symposium on Algorithms and Computation*, 21(3):405–420. DOI: 10.1007/PL00009429. 59

Choset, H., Lynch, K. M., Hutchinson, S., Kantor, G. A., Burgard, W., Kavraki, L. E., and Thrun, S. (2005). *Principles of Robot Motion: Theory, Algorithms, and Implementations*. MIT Press, Cambridge, MA. 157

Coleman, D., Şucan, I. A., Moll, M., Okada, K., and Correll, N. (2015). Experience-based planning with sparse roadmap spanners. In *IEEE International Conference on Robotics and Automation*. 160

Cormen, T. H., Leiserson, C. E., Rivest, R. L., and Stein, C. (2009). *Introduction to Algorithms, Third Edition*. The MIT Press. 13, 14, 16, 19

Daniel, K., Nash, A., Koenig, S., and Felner, A. (2010). Theta*: Any-angle path planning on grids. *Journal of Artificial Intelligence Research*, 39:533–579. DOI: 10.1613/jair.2994. 26

de Berg, M., Cheong, O., van Kreveld, M., and Overmars, M. (2008). *Computational Geometry: Algorithms and Applications*. Springer. DOI: 10.1007/978-3-540-77974-2. 5, 35, 40, 48, 58

Delling, D., Goldberg, A. V., Nowatzyk, A., and Werneck, R. F. F. (2011). Phast: Hardware-accelerated shortest path trees. In *IPDPS*, pages 921–931. DOI: 10.1109/IPDPS.2011.89. 64

Devillers, O. and Pion, S. (2003). Efficient exact geometric predicates for delaunay triangulations. In *Proceedings of the 5th Workshop Algorithm Engineering and Experiments*, pages 37–44. 60

Dijkstra, E. W. (1959). A note on two problems in connexion with graphs. *Numerische Mathematik*, 1:269–271. DOI: 10.1007/BF01386390. 13, 67

Ersson, T. and Hu, X. (2001). Path planning and navigation of mobile robots in unknown environments. In *Intelligent Robots and Systems, 2001. Proceedings. 2001 IEEE/RSJ International Conference on*, volume 2, pages 858–864 vol.2. DOI: 10.1109/IROS.2001.976276. 68

Felner, A., Zahavi, U., Holte, R., Schaeffer, J., Sturtevant, N., and Zhang, Z. (2011). Inconsistent heuristics in theory and practice. *Artificial Intelligence*, 175(9-10):1570–1603. DOI: 10.1016/j.artint.2011.02.001. 27

Ferguson, C. and Korf, R. E. (1988). Distributed tree search and its application to alpha-beta pruning. In *AAAI*, pages 128–132. 64

Fischer, L. G., Silveira, R., and Nedel, L. (2009). Gpu accelerated path-planning for multi-agents in virtual environments. In *Proccedings of the Brazilian Symposium on Games and Digital Entertainment*, SBGAMES, pages 101–110. IEEE Computer Society. DOI: 10.1109/SBGAMES.2009.20. 64

Frey, P. J. and Marechal, L. (1998). Fast adaptive quadtree mesh generation. In *Proceedings of the Seventh International Meshing Roundtable*, pages 211–224. 65

Garcia, F., Kapadia, M., and Badler, N. (2014). Gpu-based dynamic search on adaptive resolution grids. In *Robotics and Automation (ICRA), 2014 IEEE International Conference on*, pages 1631–1638. DOI: 10.1109/ICRA.2014.6907070. 101

Geisberger, R., Sanders, P., Schultes, D., and Delling, D. (2008). Contraction hierarchies: Faster and simpler hierarchical routing in road networks. In *Proceedings of the 7th International Conference on Experimental Algorithms*, WEA'08, pages 319–333, Berlin, Heidelberg. Springer-Verlag. DOI: 10.1007/978-3-540-68552-4_24. 76

Geraerts, R. (2010). Planning short paths with clearance using explicit corridors. In *ICRA'10: Proceedings of the IEEE International Conference on Robotics and Automation*. DOI: 10.1109/ROBOT.2010.5509263. 48, 59, 65

Gini, M. (1996). Parallel search algorithms for robot motion planning. In *IEEE ICRA Current Approaches and Future Directions*. DOI: 10.1109/ROBOT.1993.292122. 64

Gleicher, M., Shin, H. J., Kovar, L., and Jepsen, A. (2003). Snap-together motion: assembling run-time animations. In *Proceedings of the Symposium on Interactive 3D graphics and Games (I3D)*, pages 181–188, NY, USA. DOI: 10.1145/1201775.882333. 153

Gochev, K., Cohen, B. J., Butzke, J., Safonova, A., and Likhachev, M. (2011). Path planning with adaptive dimensionality. In Borrajo, D., Likhachev, M., and López, C. L., editors, *SOCS*. AAAI Press. 140

Goldberg, A. (2001). Shortest path algorithms: Engineering aspects. In *Proceedings of ESAAC '01, Lecture Notes in Computer Science*, pages 502–513. Springer-Verlag. DOI: 10.1007/3-540-45678-3_43. 22

Goldberg, A. V. and Harrelson, C. (2005). Computing the shortest path: A search meets graph theory. In *Proceedings of the Sixteenth Annual ACM-SIAM Symposium on Discrete Algorithms (SODA)*, pages 156–165, Philadelphia, PA, Society for Industrial and Applied Mathematics. 27

Goldenstein, S., Karavelas, M., Metaxas, D., Guibas, L., and Goswami, A. (2001). Scalable dynamical systems for multi-agent steering and simulation. In *Robotics and Automation, 2001. Proceedings 2001 ICRA. IEEE International Conference*. vol. 4, pp. 3973–3980. DOI: 10.1016/S0097-8493(01)00153-4. 65

Goodrich, M. T. and Sun, J. Z. (2005). The skip quadtree: a simple dynamic data structure for multidimensional data. In *Proc. 21st ACM Symposium on Computational Geometry*, pages 296–305. ACM. DOI: 10.1145/1064092.1064138. 65

Guy, S. J., Chhugani, J., Kim, C., Satish, N., Lin, M. C., Manocha, D., and Dubey, P. (2009). Clearpath: Highly parallel collision avoidance for multi-agent simulation. In *ACM SIGGRAPH/EUROGRAPHICS SCA*, pages 177–187. DOI: 10.1145/1599470.1599494. 64

Harabor, D. and Botea, A. (2008). Hierarchical path planning for multi-size agents in heterogeneous environments. In Philip Hingston, L. B., editor, *IEEE Symposium on Computational Intelligence and Games*, pages 258–265, Perth, Australia. 64

Hart, P., Nilsson, N., and Raphael, B. (1968). A formal basis for the heuristic determination of minimum cost paths. *IEEE Transactions on Systems Science and Cybernetics*, 4(2):100–107. DOI: 10.1109/TSSC.1968.300136. 22, 102

Haumont, D., Debeir, O., and Sillion, F. (2003). Volumetric cell-and-portal generation. In *Proceedings of EUROGRAPHICS*. DOI: 10.1111/1467-8659.00677. 56

Hauser, K. and Latombe, J.-C. (2010). Multi-modal motion planning in non-expansive spaces. *International Journal of Robotics Research*, 7(29):897–915. DOI: 10.1177/0278364909352098. 162

Helbing, D., Farkas, I., and Vicsek, T. Simulating dynamical features of escape panic. *Nature*, page 2000. DOI: 10.1038/35035023. 164

Helbing, D., Johansson, A., and Al-Abideen, H. Z. (2007). The Dynamics of crowd disasters: an empirical study. *Physical Review E*, 75(4):046109. DOI: 10.1103/PhysRevE.75.046109. 164

Helbing, D. and Mukerji, P. (2012). Crowd disasters as systemic failures: Analysis of the love parade disaster. *CoRR*, abs/1206.5856. DOI: 10.1140/epjds7. 164

Hernandez, E., Carreras, M., Galceran, E., and Ridao, P. (2011). Path planning with homotopy class constraints on bathymetric maps. In *OCEANS - Europe*. DOI: 10.1109/Oceans-Spain.2011.6003519. 65

Hershberger, J. and Snoeyink, J. (1994). Computing minimum length paths of a given homotopy class. *Computational Geometry Theory and Application*, 4(2):63–97. DOI: 10.1016/0925-7721(94)90010-8. 32

Hershberger, J. and Suri, S. (1997). An optimal algorithm for euclidean shortest paths in the plane. *SIAM Journal on Computing*, 28:2215–2256. DOI: 10.1137/S0097539795289604. 40

Hjelle, O. and Dæhlen, M. (2006). *Triangulations and Applications (Mathematics and Visualization)*. Springer-Verlag New York, Inc., Secaucus, NJ. DOI: 10.1007/3-540-33261-8. 50

Hoff III, K. E., Culver, T., Keyser, J., Lin, M., and Manocha, D. (2000). Fast computation of generalized voronoi diagrams using graphics hardware. In *ACM Symposium on Computational Geometry*. DOI: 10.1145/311535.311567. 48

Hoisie, A., Lubeck, O. M., and Wasserman, H. J. (1998). Performance analysis of wavefront algorithms on very-large scale distributed systems. In *Wide Area Networks and High Performance Computing*, pages 171–187. DOI: 10.1007/BFb0110087. 64

Huang, T., Kapadia, M., Badler, N. I., and Kallmann, M. (2014). Path planning for coherent and persistent groups. In *2014 IEEE International Conference on Robotics and Automation, ICRA 2014*, Hong Kong, China, May 31–June 7, 2014, pages 1652–1659. DOI: 10.1109/ICRA.2014.6907073. 125

Huang, Y. and Kallmann, M. (2015). Planning motions and placements for virtual demonstrators. *IEEE Transactions on Visualization and Computer Graphics (TVCG)*. DOI: 10.1109/TVCG.2015.2446494. 163

Huang, Y., Mahmudi, M., and Kallmann, M. (2011). Planning humanlike actions in blending spaces. In *Proceedings of the IEEE International Conference on Intelligent Robots and Systems (IROS)*, pages 2653–2659. DOI: 10.1109/IROS.2011.6095133. 161, 162

Jiang, X. and Kallmann, M. (2007). Learning humanoid reaching tasks in dynamic environments. In *Proceedings of the IEEE International Conference on Intelligent Robots and Systems (IROS)*, pages 1148–1153, San Diego CA. DOI: 10.1109/IROS.2007.4399542. 160

Juarez-Perez, A., Feng, A., Shapiro, A., and Kallmann, M. (2014). Deformation, parameterization and analysis of a single locomotion cycle. In *Proceedings of Motion in Games (MIG), poster abstract*. DOI: 10.1145/2668064.2677082. 164

Kaelbling, L. P. and Lozano-Pérez, T. (2011). Hierarchical planning in the now. In *IEEE Conference on Robotics and Automation (ICRA)*. Finalist, Best Manipulation Paper Award. DOI: 10.1109/ICRA.2011.5980391. 64

Kajita, S., Kanehiro, F., Kaneko, K., Yokoi, K., and Hirukawa, H. (2001). The 3d linear inverted pendulum mode: a simple modeling for a biped walking pattern generation. In *Proceedings of IEEE/RSJ (IROS)*, volume 1, pages 239–246. DOI: 10.1109/IROS.2001.973365. 131

Kallmann, M. (2005). Scalable solutions for interactive virtual humans that can manipulate objects. In *Proceedings of the Artificial Intelligence and Interactive Digital Entertainment (AI-IDE'05)*, pages 69–74, Marina del Rey, CA. 159

Kallmann, M. (2010). Shortest paths with arbitrary clearance from navigation meshes. In *Proceedings of the Eurographics / SIGGRAPH Symposium on Computer Animation (SCA)*, pages 159–168. 9

Kallmann, M. and Kapadia, M. (2014). Navigation meshes and real-time dynamic planning for virtual worlds, *Special Interest Group on Computer Graphics and Interactive Techniques Conference*, SIGGRAPH'14, Vancouver, Canada, August 10–14, 2014, Courses, 3:1–3:81, 2014. DOI: 10.1145/2614028.2615399. xv

Kallmann, M. (2014). Dynamic and robust local clearance triangulations. *ACM Transactions on Graphics*, 33(5). DOI: 10.1145/2580947. 52, 53, 54, 60, 65, 74, 90

Kallmann, M., Aubel, A., Abaci, T., and Thalmann, D. (2003). Planning collision-free reaching motions for interactive object manipulation and grasping. *Computer Graphics Forum (Proceedings of Eurographics'03)*, 22(3):313–322. DOI: 10.1111/1467-8659.00678. 159

Kallmann, M., Huang, Y., and Backman, R. (2010). A skill-based motion planning framework for humanoids. In *Proceedings of the International Conference on Robotics and Automation (ICRA)*. DOI: 10.1109/ROBOT.2010.5509939. 162

Kallmann, M. and Matarić, M. (2004). Motion planning using dynamic roadmaps. In *Proceedings of the IEEE International Conference on Robotics and Automation (ICRA)*, pages 4399–4404, New Orleans, Louisiana. DOI: 10.1109/ROBOT.2004.1302410. 158

Kapadia, M. and Badler, N. I. (2013). Navigation and steering for autonomous virtual humans. *Wiley Interdisciplinary Reviews: Cognitive Science*. DOI: 10.1002/wcs.1223. 66

Kapadia, M., Beacco, A., Garcia, F. M., Reddy, V., Pelechano, N., and Badler, N. I. (2013). Multi-domain real-time planning in dynamic environments. In *The ACM SIGGRAPH / Eurographics Symposium on Computer Animation, SCA '13*, Anaheim, CA, July 19–21, 2013, pages 115–124. DOI: 10.1145/2485895.2485909. 140, 149

Kapadia, M., Falk, J., Zünd, F., Marti, M., Sumner, R. W., and Gross, M. (2015a). Computer-assisted authoring of interactive narratives. In *Proceedings of the 19th Symposium on Interactive 3D Graphics and Games*, i3D '15, pages 85–92, New York, NY, ACM. DOI: 10.1145/2699276.2699279. 164

Kapadia, M., Garcia, F., Boatright, C., and Badler, N. (2013c). Dynamic search on the GPU. In *Intelligent Robots and Systems (IROS), 2013 IEEE/RSJ International Conference on*, pages 3332–3337. DOI: 10.1109/IROS.2013.6696830. 101

Kapadia, M., Ninomiya, K., Shoulson, A., Garcia, F., and Badler, N. I. (2013d). Constraint-aware navigation in dynamic environments. In *Proceedings of the 2013 ACM SIGGRAPH Conference on Motion in Games (MIG)*, MIG'13. DOI: 10.1145/2522628.2522654. 73, 95

Kapadia, M., Pelechano, N., Allbeck, J., and Badler, N. (2015b). Virtual crowds: Steps toward behavioral realism. *Synthesis Lectures on Visual Computing*, 7(4):1–270. DOI: 10.2200/S00673ED1V01Y201509CGR020. xv, 66, 163

Kapadia, M., Shoulson, A., Durupinar, F., and Badler, N. I. (2013e). Authoring Multi-actor Behaviors in Crowds with Diverse Personalities. In Ali, S., Nishino, K., Manocha, D., and Shah, M., editors, *Modeling, Simulation and Visual Analysis of Crowds*, volume 11 of *The Intl. Series in Video Computing*, pages 147–180. Springer New York. DOI: 10.1007/978-1-4614-8483-7. 164

Kapadia, M., Singh, S., Allen, B., Reinman, G., and Faloutsos, P. (2009a). Steerbug: an interactive framework for specifying and detecting steering behaviors. In *ACM SIGGRAPH/Eurographics Symp. on Computer Animation*, SCA '09, pages 209–216, New York, NY, ACM. DOI: 10.1145/1599470.1599497. 163

Kapadia, M., Singh, S., Hewlett, W., and Faloutsos, P. (2009b). Egocentric affordance fields in pedestrian steering. In *Proceedings of the 2009 ACM SIGGRAPH Symposium on Interactive 3D Graphics and Games*, I3D'09, pages 215–223, New York, NY, ACM. DOI: 10.1145/1507149.1507185. 66

Kapadia, M., Singh, S., Hewlett, W., Reinman, G., and Faloutsos, P. (2012). Parallelized ego-centric fields for autonomous navigation. *The Visual Computer*, pages 1–19. 10.1007/s00371-011-0669-5. DOI: 10.1007/s00371-011-0669-5. 66

Kapadia, M., Singh, S., Reinman, G., and Faloutsos, P. (2011a). A behavior-authoring framework for multiactor simulations. *Computer Graphics and Applications, IEEE*, 31(6):45–55. DOI: 10.1109/MCG.2011.68. 164

Kapadia, M., Wang, M., Reinman, G., and Faloutsos, P. (2011b). Improved benchmarking for steering algorithms. In *Motion in Games*, Lecture Notes in Computer Science, pages 266–277, Berlin, Heidelberg. Springer-Verlag. 163

Kapadia, M., Wang, M., Singh, S., Reinman, G., and Faloutsos, P. (2011c). Scenario space: Characterizing coverage, quality, and failure of steering algorithms. In *Proceedings of the 2011 Eurographics/ACM SIGGRAPH Symposium on Computer Animation, SCA 2011, Vancouver, BC, Canada, 2011*, pages 53–62. DOI: 10.1145/2019406.2019414. 149, 163

Kapadia, M., Xianghao, X., Kallmann, M., Nitti, M., Coros, S., Sumner, R. W., and Gross, M. (2016). PRECISION: Precomputed Environment Semantics for Contact-Rich Character Animation. In *Proceedings of the 2016 ACM SIGGRAPH Symposium on Interactive 3D Graphics and Games*, I3D'16, New York, NY, ACM. 163

Kapoor, S., Maheshwari, S. N., and Mitchell, J. S. B. (1997). An efficient algorithm for euclidean shortest paths among polygonal obstacles in the plane. In *Discrete and Computational Geometry*, volume 18, pages 377–383. DOI: 10.1007/PL00009323. 50, 58

Karaman, S. and Frazzoli, E. (2011). Sampling-based algorithms for optimal motion planning. *International Journal of Robotics Research*, 30(7):846–894. DOI: 10.1177/0278364911406761. 157

Kavraki, L., Svestka, P., Latombe, J.-C., and Overmars, M. (1996). Probabilistic roadmaps for fast path planning in high-dimensional configuration spaces. *IEEE Transactions on Robotics and Automation*, 12:566–580. DOI: 10.1109/70.508439. 9, 156

Kider, J., Henderson, M., Likhachev, M., and Safonova, A. (2010). High-dimensional planning on the gpu. In *ICRA*. 64

Klein, P., Rao, S., Rauch, M., and Subramanian, S. (1994). Faster shortest-path algorithms for planar graphs. *Journal of Computer and System Sciences*, pages 27–37. DOI: 10.1006/jcss.1997.1493. 27, 58

Koenig, S. and Likhachev, M. (2002). D* Lite. In *Proceedings of the 2002 National Conference on Artificial Intelligence*, pages 476–483. AAAI. 68, 69

Koenig, S., Likhachev, M., Liu, Y., and Furcy, D. (2004). Incremental heuristic search in ai. *AI Magazine*, 25(2):99–112. DOI: 10.1609/aimag.v25i2.1763. 68

Kovar, L., Gleicher, M., and Pighin, F. H. (2002). Motion graphs. *Proceedings of SIGGRAPH*, 21(3):473–482. DOI: 10.1145/566570.566605. 154

Kring, A. W., Champandard, A. J., and Samarin, N. (2010). Dhpa* and shpa*: Efficient hierarchical pathfinding in dynamic and static game worlds. In Youngblood, G. M. and Bulitko, V., editors, *AIIDE*. The AAAI Press. 65

Kuffner, J. J. and LaValle, S. M. (2000). RRT-Connect: An efficient approach to single-query path planning. In *Proceedings of IEEE International Conference on Robotics and Automation (ICRA)*, San Francisco, CA. DOI: 10.1109/ROBOT.2000.844730. 10, 158

Kuo, A. (2007). The six determinants of gait and the inverted pendulum analogy: A dynamic walking perspective. *Human Movement Science*, 26(4):617–656. European Workshop on Movement Science 2007. DOI: 10.1016/j.humov.2007.04.003. 135

Lamarche, F. (2009). Topoplan: a topological path planner for real time human navigation under floor and ceiling constraints. *Computer Graphics Forum*, 28(2):649–658. DOI: 10.1111/j.1467-8659.2009.01405.x. 56

Latombe, J.-C. (1990). *Robot Motion Planning*. Kluwer Academic Publisher. 157

LaValle, S. M. (1998) *Rapidly-Exploring Random Trees: A New Tool for Path Planning*, Iowa State University, Computer Science Department, 98-11, October. 157

LaValle, S. M. (2006). *Planning Algorithms*. Cambridge University Press (available on-line). DOI: 10.1017/CBO9780511546877. 13, 130, 157

Lee, D. T. and Preparata, F. P. (1984). Euclidean shortest paths in the presence of rectilinear barriers. *Networks*, 3(14):393–410. DOI: 10.1002/net.3230140304. 32

Li, S.-X. and Loew, M. H. (1987a). Adjacency detection using quadcodes. *Communications of ACM*, 30(7):627–631. DOI: 10.1145/28569.28574. 65, 114, 115

Li, S.-X. and Loew, M. H. (1987b). The quadcode and its arithmetic. *Communications of ACM*, 30(7):621–626. DOI: 10.1145/28569.28573. 65, 115

Likhachev, M., Ferguson, D. I., Gordon, G. J., Stentz, A., and Thrun, S. (2005). Anytime Dynamic A*: An anytime, replanning algorithm. In *Proceedings of the 2005 International Conference on Automated Planning and Scheduling*, pages 262–271. 66, 68, 69, 76, 102

Likhachev, M., Gordon, G. J., and Thrun, S. (2003). Ara*: Anytime A* with provable bounds on sub-optimality. In *Proceedings of the 2003 Conference on Advances in Neural Information Processing Systems*. 67

Liu, Y. H. and Arimoto, S. (1995). Finding the shortest path of a disk among polygonal obstacles using a radius-independent graph. *IEEE Transactions on Robotics and Automation*, 11(5):682–691. DOI: 10.1109/70.466615. 35

Lozano-Pérez, T. and Wesley, M. A. (1979). An algorithm for planning collision-free paths among polyhedral obstacles. *Communications of ACM*, 22(10):560–570. DOI: 10.1145/359156.359164. 35, 58

Mahmudi, M. and Kallmann, M. (2012). Precomputed motion maps for unstructured motion capture. In *Eurographics/SIGGRAPH Symposium on Computer Animation (SCA)*. 154

Mahmudi, M. and Kallmann, M. (2013). Analyzing locomotion synthesis with feature-based motion graphs. *IEEE Transactions on Visualization and Computer Graphics*, 19(5):774–786. DOI: 10.1109/TVCG.2012.149. 154

Mahmudi, M. and Kallmann, M. (2015). Multi-modal data-driven motion planning and synthesis. In *Proceedings of the 8th International Conference on Motion In Games (MIG)*. DOI: 10.1145/2822013.2822044. 163

Massive Software Inc. (2010). Massive: Simulating life. http:www.massivesofware.com. 164

Mitchell, J. S. B. (1991). A new algorithm for shortest paths among obstacles in the plane. *Annals of Mathematics and Artificial Intelligence*, 3:83–105. DOI: 10.1007/BF01530888. 40

Mitchell, J. S. B. (1993). Shortest paths among obstacles in the plane. In *Proceedings of the Ninth Annual Symposium on Computational Geometry (SoCG)*, pages 308–317, New York, NY, ACM. DOI: 10.1145/160985.161156. 40, 41

Mitchell, J. S. B. and Papadimitriou, C. H. (1991). The weighted region problem: Finding shortest paths through a weighted planar subdivision. *Journal of the ACM*, 38(1):18–73. DOI: 10.1145/102782.102784. 43, 66, 83

Mononen, M. (2015). Recast navigation mesh toolset. https://github.com/memononen/recastnavigation. 56, 57, 74, 90

Narayanappa, S., Vojtêchovský, P., and Bae, W. D. (2005). Exact solutions for simple weighted region problems. 66, 83

Nilsson, N. (1969). A mobile automaton: an application of artificial intelligence techniques. In *Proceedings of the 1969 International Joint Conference on Artificial Intelligence (IJCAI)*, pages 509–520. 35, 58

Ninomiya, K., Kapadia, M., Shoulson, A., Garcia, F., and Badler, N. (2014). Planning approaches to constraint-aware navigation in dynamic environments. *Computer Animation and Virtual Worlds*, pages n/a–n/a. DOI: 10.1002/cav.1622. 73

Oliva, R. and Pelechano, N. (2011). Automatic generation of suboptimal navmeshes. In *Proceedings of the 2011 International Conference on Motion in Games (MIG)*, MIG'11, pages 328–339, Berlin, Heidelberg. Springer-Verlag. DOI: 10.1007/978-3-642-25090-3_28. 58, 74, 90

Oliva, R. and Pelechano, N. (2013). NEOGEN: Near optimal generator of navigation meshes for 3D multi-layered environments. *Computer & Graphics*, 37(5):403–412. DOI: 10.1016/j.cag.2013.03.004. 56, 57, 59, 74, 90

Pal, A., Tiwari, R., and Shukla, A. (2011). A focused wave front algorithm for mobile robot path planning. In *HAIS (1)*, pages 190–197. DOI: 10.1007/978-3-642-21219-2_25. 64, 102

Park, S. I., Shin, H. J., Kim, T. H., and Shin, S. Y. (2004). On-line motion blending for real-time locomotion generation. *Computer Animation and Virtual Worlds*, 15(3-4):125–138. DOI: 10.1002/cav.15. 153

Pelechano, N., Allbeck, J. M., and Badler, N. I. (2008). *Virtual Crowds: Methods, Simulation, and Control*. Synthesis Lectures on Computer Graphics and Animation. Morgan & Claypool Publishers. DOI: 10.2200/S00123ED1V01Y200808CGR008. xv, 66

Pettré, J., Laumond, J.-P., and Thalmann, D. (2005). A navigation graph for real-time crowd animation on multilayered and uneven terrain. *First International Workshop on Crowd Simulation*, 43(44):194. 74, 90

Phillips, M., Hwang, V., Chitta, S., and Likhachev, M. (2013). Learning to plan for constrained manipulation from demonstrations. DOI: 10.1007/s10514-015-9440-5. 65

Ira Pohl. (1973). The avoidance of (relative) catastrophe, heuristic competence, genuine dynamic weighting and computational issues in heuristic problem solving. in *Proceedings of the 3rd International Joint Conference on Artificial Intelligence*, p. 12–17, 1973. 67

Powley, C., Ferguson, C., and Korf, R. E. (1993). Depth-first heuristic search on a simd machine. *Artificial Intelligence*, 60(2):199–242. DOI: 10.1016/0004-3702(93)90002-S. 64

Preparata, F. P. and Shamos, M. I. (1985). *Computational Geometry - An Introduction*. Springer. DOI: 10.1007/978-1-4612-1098-6. 48

Ramalingam, G. and Reps, T. (1996). An incremental algorithm for a generalization of the shortest-path problem. *Journal of Algorithms*, 21(2):267–305. DOI: 10.1006/jagm.1996.0046. 68

Rayner, C., Bowling, M., and Sturtevant, N. (2011). Euclidean heuristic optimization. In *Proceedings of the Twenty-Fifth Conference on Artificial Intelligence (AAAI)*, pages 81–86. 27

Reif, J. and Sun, Z. (2000). An efficient approximation algorithm for weighted region shortest path problem. 66, 83

Rokicki, T., Kociemba, H., Davidson, M., and Dethridge, J. (2013). The diameter of the rubik's cube group is twenty. *SIAM Journal on Discrete Mathematics*, 27(2):1082–1105. DOI: 10.1137/120867366. 13

Schultes, D. (2008). *Route Planning in Road Networks*. VDM Verlag, Saarbrücken, Germany. 76

Schwarzer, F., Saha, M., and Latombe, J.-C. (2002). Exact collision checking of robot paths. In *Proceedings of the Workshop on Algorithmic Foundations of Robotics (WAFR'02)*, Nice. DOI: 10.1007/978-3-540-45058-0_3. 159, 160

Shewchuk, J. R. (1997). Adaptive precision floating-point arithmetic and fast robust geometric predicates. *Discrete & Computational Geometry*, 18(3):305–363. DOI: 10.1007/PL00009321. 60

Shimoda, S., Kuroda, Y., and Iagnemma, K. (2005). Potential field navigation of high speed unmanned ground vehicles on uneven terrain. *Proceedings of the 2005 IEEE International Conference on Robotics and Automation*, pages 2828–2833. DOI: 10.1109/ROBOT.2005.1570542. 65

Shoulson, A., Gilbert, M. L., Kapadia, M., and Badler, N. I. (2013a). An event-centric planning approach for dynamic real-time narrative. In *Proceedings of Motion on Games*, MIG '13, pages 99:121–99:130, New York, NY, ACM. DOI: 10.1145/2522628.2522629. 164

Shoulson, A., Marshak, N., Kapadia, M., and Badler, N. I. (2013b). ADAPT: The agent development and prototyping testbed. In *ACM SIGGRAPH Symp. on Interactive 3D Graphics and Games*, I3D '13, pages 9–18, New York, NY, ACM. DOI: 10.1145/2448196.2448198. 87, 164

Shoulson, A., Marshak, N., Kapadia, M., and Badler, N. I. (2014a). ADAPT: The agent developmentand prototyping testbed. *IEEE Transactions on Visualization and Computer Graphics*, 20(7):1035–1047. DOI: 10.1109/TVCG.2013.251. 148, 164

Singh, S., Kapadia, M., Faloutsos, P., and Reinman, G. (2009b). SteerBench: a benchmark suite for evaluating steering behaviors. *Computer Animation and Virtual Worlds*, 9999(9999). DOI: 10.1002/cav.277. 149, 163

Singh, S., Kapadia, M., Reinman, G., and Faloutsos, P. (2011a). Footstep navigation for dynamic crowds. *Computer Animation and Virtual Worlds*, 22(2-3):151–158. DOI: 10.1145/1944745.1944783. 131

Singh, S., Kapadia, M., Reinman, G., and Faloutsos, P. (2011b). Footstep navigation for dynamic crowds. In *Symposium on Interactive 3D Graphics and Games*, I3D '11, San Francisco, CA, February 18–20, 2011, page 203. DOI: 10.1145/1944745.1944783. 131

Singh, S., Naik, M., Kapadia, M., Faloutsos, P., and Reinman, G. (2008). Watch out! a framework for evaluating steering behaviors. In *Motion in Games*, Lecture Notes in Computer Science, pages 200–209, Berlin, Heidelberg. Springer-Verlag. DOI: 10.1007/978-3-540-89220-5_20. 163

Snook, G. (2000). Simplified 3d movement and pathfinding using navigation meshes. In De-Loura, M., editor, *Game Programming Gems*, pages 288–304. Charles River Media. 47

Stentz, A. (1995). The focussed d* algorithm for real-time replanning. In *Proceedings of the 14th International Joint Conference on Artificial Intelligence - Volume 2*, IJCAI'95, pages 1652–1659, San Francisco, CA. Morgan Kaufmann Publishers Inc. 68

Sturtevant, N. R. (2007). Memory-efficient abstractions for pathfinding. In Schaeffer, J. and Mateas, M., editors, *AIIDE*, pages 31–36. The AAAI Press. 64

Sturtevant, N. R. (2009). Optimizing motion-constrained pathfinding. In Darken, C. and Youngblood, G. M., editors, *AIIDE*. The AAAI Press. 65

Sturtevant, N. R. (2012). Benchmarks for grid-based pathfinding. *Transactions on Computational Intelligence and AI in Games*, 4(2):144–148. DOI: 10.1109/TCIAIG.2012.2197681. 94, 107, 108, 109

Sturtevant, N. R. (2013). Incorporating human relationships into path planning. In *Proceedings of the 9th AAAI Conference on Artificial Intelligence and Interactive Digital Entertainment*. 65, 81, 95

Şucan, I. A., Moll, M., and Kavraki, L. E. (2012). The Open Motion Planning Library. *IEEE Robotics & Automation Magazine*, 19(4):72–82. http://ompl.kavrakilab.org. DOI: 10.1109/MRA.2012.2205651. 160

Thalmann, D. and Musse, S. R. (2007). *Crowd Simulation*. Springer. DOI: 10.1007/978-1-4471-4450-2. xv, 66

The CGAL Project (2014). *CGAL User and Reference Manual*. CGAL Editorial Board, 4.4 edition. http://doc.cgal.org/4.4/Manual/packages.html. 50

Torchelsen, R. P., Scheidegger, L. F., Oliveira, G. N., Bastos, R., and Comba, J. A. L. D. (2010). Real-time multi-agent path planning on arbitrary surfaces. In *Proc. ACM SIGGRAPH symposium on Interactive 3D Graphics and Games*, I3D, pages 47–54. DOI: 10.1145/1730804.1730813. 64

Tozour, P. (2002). Building a near-optimal navigation mesh. In Rabin, S., editor, *AI Game Programming Wisdom*, pages 171–185. Charles River Media. 47

Treuille, A., Cooper, S., and Popović, Z. (2006). Continuum crowds. *ACM Transactions on Graphics*, 25(3):1160–1168. DOI: 10.1145/1141911.1142008. 66

Tripath Toolkit (2010). Triangulation and path planning toolkit. http://graphics.ucmerced.edu/software/tripath/. 50

van Toll, W. G., Cook, A. F., and Geraerts, R. (2012). A navigation mesh for dynamic environments. *Computer Animation and Virtual Worlds*, 23(6):535–546. DOI: 10.1002/cav.1468. 74, 90

van Toll, W. G., IV, A. F. C., and Geraerts, R. (2011). Navigation meshes for realistic multi-layered environments. In *Proceedings of the IEEE/RSJ International Conference on Intelligent Robots and Systems (IROS)*, pages 3526–3532. DOI: 10.1109/IROS.2011.6094790. 57

Warren, C. (1989). Global path planning using artificial potential fields. In *Proceedings of the 1989 IEEE International Conference on Robotics and Automation*, volume 1, pages 316–321. DOI: 10.1109/ROBOT.1989.100007. 65

Warren, C. (1990). Multiple robot path coordination using artificial potential fields. In *Proceedings of the 1990 IEEE International Conference on Robotics and Automation*, volume 1, pages 500–505. DOI: 10.1109/ROBOT.1990.126028. 65

Wein, R., van den Berg, J., and Halperin, D. (2007). The visibility-voronoi complex and its applications. *Computational Geometry: Theory and Applications*, 36(1):66–78. DOI: 10.1016/j.comgeo.2005.11.007. 35, 45

Xu, Y. D. and Badler, N. I. (2000). Algorithms for generating motion trajectories described by prepositions. In *Proceedings of Computer Animation 2000*, pages 30–35. DOI: 10.1109/CA.2000.889029. 65

Zhou, Y. and Zeng, J. (2015). Massively parallel a* search on a gpu. 64

Authors' Biographies

MARCELO KALLMANN

Marcelo Kallmann is Founding Faculty and Associate Professor of Computer Science at the School of Engineering of the University of California, Merced. He holds a Ph.D. from the Swiss Federal Institute of Technology in Lausanne (EPFL), was Research Faculty at the University of Southern California (USC), and a scientist at the USC Institute for Creative Technologies (ICT) before moving to UC Merced in 2005. His areas of research include computer animation, virtual reality, and motion planning. At UC Merced, he established and leads the computer graphics research group. His research work has been supported by several awards from the US National Science Foundation, and his work on triangulations for path planning runs inside The Sims 4, the latest installment of one of the best-selling video game series of all time.

MUBBASIR KAPADIA

Mubbasir Kapadia is an Assistant Professor in the Computer Science Department at Rutgers University. Previously, he was an Associate Research Scientist at Disney Research Zurich. He was a postdoctoral researcher and Assistant Director at the Center for Human Modeling and Simulation at University of Pennsylvania, under the directorship of Prof. Norman I. Badler. He was the project lead on the United States Army Research Laboratory (ARL) funded project Robotics Collaborative Technology Alliance (RCTA). He received his Ph.D. in computer science at University of California, Los Angeles under the advisement of Professor Petros Faloutsos. He is the co-author of the book *Virtual Crowds: Steps Toward Behavioral Realism*, Morgan & Claypool Publishers, 2015.